THE OPEN CIRCLE

MARTIN L. COOK

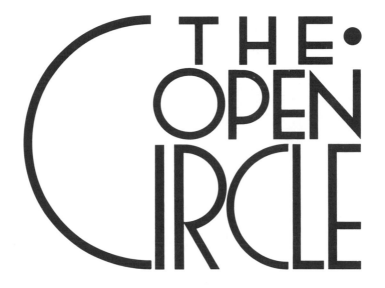

THE·OPEN CIRCLE

CONFESSIONAL

METHOD IN

THEOLOGY

FORTRESS PRESS • MINNEAPOLIS

THE OPEN CIRCLE
Confessional Method in Theology

Cover design: Brian Preuss

Library of Congress Cataloging-in-Publication Data

Cook, Martin L., 1951–
    The open circle : confessional method in theology / Martin L. Cook.
        p.   cm.
    Includes bibliographical references and index.
    ISBN 0-8006-2482-3 (alk. paper)
    1. Theology—Methodology.   2. Knowledge, Theory of (Religion)
3. Science—Philosophy.   4. Niebuhr, H. Richard (Helmut Richard),
1894–1962.   I. Title.
BR118.C67   1991
230'.01—dc20   91-14230
                                                                                      CIP

The paper used in this publication meets the minimum requirements of American National Standard for Information Sciences—Permanence of Paper for Printed Library Materials, ANSI Z329.48-1984.   ∞ ™

Manufactured in the U.S.A.   AF 1-2482

95   94   93   92   91   1   2   3   4   5   6   7   8   9   10

"You propose then, Philo, said Cleanthes, to erect religious faith on philosophical scepticism; and you think that if certainty or evidence be expelled from every other subject of enquiry, it will all retire to these theological doctrines, and there acquire a superior force and authority. Whether your scepticism be as absolute and sincere as you pretend, we shall learn bye and bye. . . . We shall then see, whether you go out at the door or the window; and whether you really doubt, if your body has gravity, or can be injured by its fall."

David Hume, *Dialogues Concerning Natural Religion,* Part I

CONTENTS

PREFACE

Religion, among other things, provides an interpretation of the world, its history, and the place of humanity in it. In that respect, religious thought (perhaps especially Christian thought) purports to give a comprehensive and true account of things in the world.

Yet, precisely this claim gives rise to problems: the relativity and "cognitive minority status" of religious communities, and the ambiguous character that that relativity imparts to their thought. Because of that relativity, all vital religious communities risk becoming intellectually isolated from the intellectual developments of their time and place. This can happen either inadvertently or, more commonly, by a community's deliberate efforts to isolate itself and preserve the purity of its traditional theological formulations. Like most contemporary academic theologians, I sense the danger such isolation represents to the personal and intellectual integrity of the Christian community and to individual Christians. In articulating an understanding of theological method, I want to ensure that theology not become insular and parochial.

An equally vital concern, in my judgment, has not been given adequate attention: Whatever safeguards against unnecessary intellectual isolation a theological method includes, theology that accurately expresses the convictions of a vital religious community needs to go beyond whatever is completely intelligible and persuasive to all members of the theologian's culture. To put the matter in Pauline language, all theology that articulates true religious conviction will possess a "scandal" (1 Cor. 1:23)—a point at which religious conviction transcends the sweetly reasonable.

This book, then, attempts to steer a course between unacceptable extremes. Theology is scandalous because it is, indeed, must be, confessional. It emerges from convictions that derive from the inevitable particularity of distinctive religious history and perception. It must, however, pay adequate attention to this necessary truth without providing carte blanche for religious thought to depart wholly from the legitimate intellectual demands of other intellectual disciplines in its time and culture.

This book, however, is a "philosophy of theology," and not theology proper. It does not examine the adequacy of a theological method demonstrating substantive results that are recognized as such by the relevant community. It constitutes only a prolegomenon to theology, the adequacy of which will be demonstrated only when substantive work bears fruit that is acknowledged as a contribution to the self-understanding of the Christian community.

Nevertheless, an accounting of our present situation provides a key. In Western history, religious explanation once was thoughtful people's predominant source of "truth." But theology no longer plays that role; for most people in our culture, science and its technological progeny have usurped that dominant, if not exclusive, role in providing definitive explanations of the world in which we live. When science emerged in the West, head-to-head conflicts occurred between representatives of the two perspectives, as with Galileo and Darwin. From those conflicts, thoughtful religion learned a painful but valuable lesson: Something fundamental was mistaken in a tactic that attempted confrontation between science and religion on the same plane.

Recent developments in the philosophy of science have reopened the question of the relationship between scientific thought and theology. The historicist philosophy of science associated most commonly with Thomas Kuhn challenges an older understanding of scientific method as simply transcribing the facts of the world off the face of nature, and has stressed the sociological character of disciplines and the historically evolutionary character of scientific concepts and disciplines.

PREFACE

Unfortunately, theologians often embrace this new approach to science as supportive of unbridled relativism in truth-claims. Some theologians have declared that because even scientific claims are relative, they have no place of privilege in comparison to religious ones, while any religious claim is accepted as long as some community uses it and judges it to be true. This book explores this issue on the following terms: What, precisely, are the claims of the new philosophy of science? What are the uses and misuses of the important stresses on historicism among theologians? Most importantly, what implications follow from this discussion for the proper understanding of constructive theological method?

My writing of this book has been greatly facilitated and generously supported by the Department of Religious Studies and the College of Arts and Science of Santa Clara University. I am also greatly indebted to those who criticized earlier drafts. Langdon Gilkey and David Tracy provided guidance and criticism of a doctoral dissertation that was the seed of this book. Douglas Ottati has consistently encouraged the development of these ideas. J. Michael West, Fortress Press, provided invaluable suggestions for revision and restructure of the manuscript. James M. Gustafson, to whom I am especially indebted, both intellectually and personally, exemplified relentless intellectual honesty and clarity that I strive to emulate.

ABBREVIATIONS

CD   Barth, Karl. *Church Dogmatics*. 2 vols. Edited by G. W. Bromiley and T. F. Torrance. Translated by G. W. Bromiley et al. Edinburgh: T. & T. Clark, 1975.

CP   Phillips, D. Z. *The Concept of Prayer*. London: Routledge and Kegan Paul, 1965.

DGE   Phillips, D. Z. "Does God Exist? In *RWE*.

FPE   Phillips, D. Z. *Faith and Philosophical Enquiry*. New York: Schocken Books, 1970.

FSRU   Phillips, D. Z. "Faith, Scepticism, and Religious Understanding." In *FPE*.

FT   Fiorenza, Francis Schüssler. *Foundational Theology: Jesus and the Church*. New York: Crossroad, 1984.

GF   Holmer, Paul. *The Grammar of Faith*. New York: Harper & Row, 1978.

HU   Toulmin, Stephen. *Human Understanding*. Vol. 1. Princeton, N. J.: Princeton University Press, 1972.

INTRODUCTION

Gordon Kaufman accurately characterizes the contemporary theological scene when he notes: "That the contemporary theological scene has become chaotic is evident to everyone who attempts to work in theology. There appears to be no consensus on what the task of theology is or how theology is to be pursued."[1] Although the reasons for this chaos are legion, uncertainty about theological method—the understanding of how theology and how the theologian work—is fundamental. Various methods propose that theology attend (at least explicitly) to different sources for theological information and that sources be weighted in some relation to one another.[2] Opinions differ as to establishing the proper milieu for theologians: some see theologians as related to and primarily in dialogue with the community of university scholars; some with the community of theologians from across the spectrum of ecumenical Christian theology; and some with the particular confessional or denominational tradition in which the theologian stands.[3] Methods also differ in how one is led to understand revelation or Scripture and to employ it. These examples only scratch the surface of the plurality.

George A. Lindbeck's *The Nature of Doctrine: Theology in a Post-liberal Age*[4] has recently provided new impetus and focus to the debate over methodology. Much of what Lindbeck advocates in this work has, in broad terms and in other language, been previously espoused

1

by a variety of thinkers who can be collectively labeled "confessional theologians."[5] In general, these thinkers are united in their attack on *foundationalism,* the belief that the meaning and truth of religious intellectual schemes must be supported and warranted by showing their coherence with perspectives and information that are not distinctively religious or tied to a particular religious community. In contrast, *confessionalism* argues that theology derives its core insights and starting point from the perspective unique to the Christian religious community. On the basis of this claim, confessional thinkers reject the belief that some allegedly common human experience authorizes or justifies distinctively Christian discourse.

These thinkers also share a common theoretical difficulty: explaining how the inner perspectives of a confessional heritage interact with the outer view of other intellectual disciplines and questions. In practical terms, each must offer some account of how theology appropriately strikes a balance between accommodating changing beliefs and assumptions in the surrounding culture, on the one hand, and remaining loyal to a received tradition, on the other. Each is at least in principle required to offer some normative account of the descriptively obvious fact that theological convictions and doctrines do change through time, often in response to intellectual and cultural currents from outside the religious and theological community.

This conversation about the epistemological basis of the theological enterprise and the degree to which its statements are capable of bearing claims of truth is not unique to theology. Indeed, such questions arise through all areas of contemporary academic life, from debates among literary scholars regarding critical theory and the social location of texts, to the philosophy-of-science question of the nature of the truth-claims made by scientific theories.

A recognition is growing from this whole range of perspectives that traditional empiricist accounts of epistemology are fundamentally flawed. Whether one approaches the question of knowledge from reflection on natural science, sociology and anthropology, literary criticism and hermeneutics, or theological method, philosophical epistemology, especially that of Anglo-American Empiricists, is inadequate to describe and account for what thinkers actually do when they reason and research. Furthermore, misguided belief in that epistemology introduced artificially sharp divisions between types of reasoning and

INTRODUCTION

fields of inquiry, and those divisions have hampered efforts to understand accurately the nature of human reason. Under the sway of these models, the natural sciences have long been seen as the paradigm case of knowledge founded directly on empirical data. Their theoretical constructs were thought to have privileged place in comparison to those of other fields because of science's direct verification or falsification in experience. They alone were thought to represent knowledge in the strictest sense.[6]

In recent decades, intense analysis of the nature of reasoning and intellectual growth in the natural sciences has developed. From it a consensus is now emerging as to the proper analysis and description of scientific rationality.[7] As one perceptive commentator remarked, "Although there are many disagreements among advocates of the new approach, there are enough common themes to justify talking about a 'new image of science.' "[8] The new philosophy of science stresses the function of tradition, the place of nonrational elements in scientific thought and inquiry, and the degree to which existing theory guides and forms perception.[9] In philosophy, the legacy of the later work of Ludwig Wittgenstein gives greater attention to the social character of language, and the relation of language to the formation of perception. The critique of the attempt to found empirical knowledge in incorrigible sensory data, as well as a growing rapprochement between Anglo-American and continental philosophical traditions, has radically altered the discussion of epistemology in recent years.[10] Some thinkers have broadened the discussion to include an awareness of the fundamentally hermeneutical character of human existence itself and the degree to which ongoing interpretation is an essential element of human experience.[11] Thus, for Logical Positivists to sharply distinguish between natural science and logic, on the one hand, and all other realms of discourse, on the other, is indefensible. To hope to settle a priori the distinction between objective knowledge and all other modes of human reflection is an illusory goal. Instead, it is time to recast radically the traditional language and distinctions of epistemology and to move "beyond objectivism and relativism"[12] entirely.

This discussion has obvious implications both for a philosophical account of the nature of religious thought and for the work of constructive theology. Indeed, contemporary theology is mired in methodological debate parallel to, but often inadequately informed by, the

3

same issues that preoccupy other areas of inquiry. Much of this debate has been carried on intratheologically or in dialogue with a single non-theological perspective, but the issues raised must be situated within the framework of an overarching and cross-disciplinary discussion of the very nature of rationality.

This range of issues is illuminated in the following chapters. The goals of this analysis are to provide a *philosophy of theology*—an account of the intellectual structure of theological reasoning—and to show that some accounts of confessional theological method parallel quite closely the broad features of the emerging consensus regarding the nature of rationality in general. A variety of broadly confessional theological methods are surveyed, and the resources of the broader discussion of epistemology are applied to their differences. Ultimately, some forms of confessional method which are sufficiently nuanced and subtle in their understanding of epistemological issues will emerge to offer viable approaches to contemporary constructive theology.

Chapter 1 examines the claims of the new philosophy of science. This highly focused and precise discussion regarding natural science will greatly clarify the parallel issues in discussions of theological method, despite the much looser character of theological reasoning. Even in light of the challenges to the older positivist idea of the absolute discontinuity and superiority of the natural sciences among the ways of knowing true things about the world, the sciences provide the paradigm case of what human knowledge is when it is disciplined and careful. Science clearly progresses, each generation building on the work of prior contributors. The fact that it does so warrants careful attention to the conditions for its successes.

Changes in the way we understand science's progress require a fundamental recasting of our understanding of human knowledge and inquiry. A reconsideration of the paradigm case of knowledge will yield an outline of a new epistemological consensus that can inform and guide a treatment of the kind of knowing theology makes possible. Further, some of the fundamental truths any adequate contemporary epistemology must take into account will guide our analysis and criticism of a range of recent proposals in constructive theology as well as provide the basis of a more adequate proposal for future constructive theology.

INTRODUCTION

Following the discussion of the philosophy of science, chapter 2 surveys a spectrum or range of confessional theologians, noting their explicit and implicit epistemological claims, and critically assessing the adequacy of their constructive and normative claims. In particular, confessional theology, while having significant similarities to George A. Lindbeck's "cultural-linguistic model," will be shown to need further development in all areas of his typology: "experiential-expressive," "dogmatic-propositional," and "cultural-linguistic."[13]

Thus, chapter 3 will explore the theology of H. R. Niebuhr's particular version of confessionalism and show that he utilizes a method that provides resources for further extension and development. The understanding of theology he provides is creative and faithful to the central concerns of confessional thinkers and epistemologically adequate and open to the insights of other disciplines.

Finally, chapter 4 explores the possibilities and limitations of the analogy between religious thought confessionally understood, and the more tightly organized thought of the natural sciences. The goal is to show that some forms of confessional method are sufficiently nuanced and subtle in their understanding of epistemological issues to warrant serious consideration as viable approaches to contemporary constructive theology. The two central issues are: (1) the ways in which theology is like and unlike a scientific discipline, and therefore subject to the same sort of philosophical analysis; and (2) the ways and extent to which the analogies that do exist suggest ways of clarifying or extending the fundamental insights and methods of confessional method in theology.

My goal in this project is not to provide one more contribution to a seemingly interminable and abstract discussion of theological method in general, but rather to assess methodologically self-conscious confessional theologies that engage in substantive theological construction. The purpose is to develop a model for constructive theology that is true to both the core of inherited tradition and our best understanding of the nature of disciplined thought and inquiry in general.

1 | THE CONFESSIONAL NATURE OF SCIENCE

A NALYSIS OF THE METHODS AND presuppositions of scientific inquiry has been a major preoccupation of philosophy in this century. Scientific thought is unique in its ability to progress rapidly through the cumulative efforts of many independent researchers who are united in a common disciplinary inquiry and yet free from formal centralized control or supervision. That fact calls for a philosophical account of scientific thought.

A discussion of the philosophy of science in this century can be divided into two periods or phases. The first began with the Logical Positivism of the Vienna Circle. It originally claimed that only those propositions that could be verified by experience were meaningful. In the face of evolving philosophical criticism it retreated from this strong claim to the view that only propositions that can, in principle, be *falsified* by direct experience are meaningful. Such changes not withstanding, however, thinkers during this period were united in their attempt to specify a priori the features that distinguish scientific thought from other kinds of speech and inquiry, and to claim a unique epistemic value for scientific knowledge. Their thinking may be generally designated as Logical Empiricism.[1]

The second phase of modern philosophy of science grew out of a detailed analysis of historical examples of the actual workings of the sciences. Consequently, its representatives are generally classed as historicist philosophers of science. Their method is concisely expressed by Norwood Hanson, one of the pioneers of this approach: "Profitable philosophical discussion of any science depends on a thorough familiarity with its history and present state."[2]

The many variations of detail among the Logical Empiricist group are unified in at least two underlying assumptions: (1) a Humean theory of perception, (2) an identification of rationality with the formal deductive logic of the *Principia Mathematica* of Russell and Whitehead.

The Humean theory of perception presumes that perception consists of simple impressions that are passively received by the mind and provide the "raw data" of cognition. These impressions are then assembled by cognitive processes into a hierarchy of propositions and theories. The epistemological priority, however, always remains with the immediate percepts, which are believed to be unconditioned in any way by the constructs placed upon them and which provide the grounding and continual test for their adequacy.

The assumption of the identification of rationality with *Principia* logic causes this tradition to assume that all scientific statements must be formulable in its particular notation. On this view, the rationality of science consists in the derivation by formally logical means of propositions that can be deduced from incorrigible and indubitable empirical observations, or "observation statements."

These underlying assumptions conjointly defined the problems that preoccupied the philosophy of science during that period. Essentially, two activities predominated: "analysis of the confirmation relation that is to hold between a scientific law and the observation statements which confirm or disconfirm it, and the analysis of how scientific terms get their meaning" (*PTC*, 23–24). The problem of meaning arises because many crucial scientific terms, for example, "electron," do not refer to entities that are directly observable, and yet, on the epistemological assumptions underlying this tradition, only observables are "real."

Philosophy of science as dominated by these presuppositions failed to resolve these questions. It could not show that the actual working of scientific rationality was to be described in these terms. Its repeated failures to account for scientific thought in these terms led to the final rejection of the assumptions and a dissolution of the idea that these problems were the appropriate focus of philosophical inquiry in the first place. The failed assumptions were replaced with a new model to account for scientific thought (*PTC*, 1–77).

The decline of the Logical Empiricist tradition of the philosophy of science exemplifies a failed attempt to bring a predetermined theoretical framework to bear on a range of empirical data. That framework was inadequate to explain the data of the actual working of scientific thinking. In other words, the failure of Logical Empiricism, and its subsequent replacement with a historicist philosophy of science, is an example of a Kuhnian "paradigm shift" in its own right.[3]

Significant contemporary philosophers of science continue to have "one foot" in the assumptions of the Logical Empiricist model—most notably Karl Popper and his disciples. Nevertheless, it is indisputable that a new consensus has emerged which fundamentally challenges the presuppositions and methods of this tradition. The new consensus agrees that philosophy of science must be engaged at the deepest level with the actual course of the historical development of science,[4] and that the result of such historical analysis necessitates a radical departure from the theory of perception and understanding of rationality which provided the shared framework for Logical Empiricism.

It is this new phase in the philosophy of science that is of major concern here. Besides providing a more accurate description of the actual working of scientific rationality, it also provides numerous parallels to the theological issues at stake among confessional theologians. Those close parallels suggest that the debate within philosophy of science may provide promising possibilities for clarification of the theological debate as well.

THE "NEW" PHILOSOPHY OF SCIENCE

There are many points on which philosophers of science such as Thomas Kuhn, Stephen Toulmin, and Norwood Hanson disagree. To the extent, however, of the major points on which they agree, they collectively represent a departure from the philosophy of science which dominated the earlier portion of this century. These broad areas of agreement come under two headings: (1) the understanding of perception which they share (a departure from the broadly Humean assumptions noted above), and (2) their emphasis on the sociological and historical character of scientific disciplines and of explanation (in contrast to the a priori, abstract philosophical definitions of those and related terms which dominated the earlier debate).

One major line of attack on the older understanding of science is to reject the theory of perception which underlay it. As noted above, the Logical Empiricists assumed that the mind is passive in perception, that the "raw material" for scientific theories and the solid ground for their claims to truth consist of their correspondence to raw and uninterpreted sensory data, and that knowledge consists of a hierarchical structure "built up" from such "foundations." C. I. Lewis expressed this claim:

> Our empirical knowledge rises as a structure of enormous complexity, most parts of which are stabilized in measure by their mutual support, but all of which rest, at bottom, on direct findings of sense. Unless there should be some statements, or rather something apprehensible and statable, whose truth is determined by given experience and is not determinable in any other way, there would be no non-analytic affirmation whose truth could be determined at all, and no such thing as empirical knowledge.[5]

The new philosophy of science attacks this theory of perception and, by contrast, emphasizes the "theory-laden" character of even the most elementary kinds of observation.[6] Harold I. Brown summarizes this development:

> One of the cornerstones of logical empiricism is the thesis that there is a fundamental distinction between uninterpreted scientific theories and the body of perceptual experience which confers meaning on our theories and determines which ones are to be accepted. . . . In response to the view that perception provides us with pure facts, it is argued that the knowledge, beliefs and theories we already hold play a fundamental role in determining what we perceive. (*PTC*, 81)

These philosophers argue that perception is not of raw and uninterpreted sensory data. Drawing on experimental psychology, especially gestalt shift experiments in which one drawing or painting can be seen as one of two different images with abrupt transitions and discontinuities between the perceptions,[7] they point out the large contribution the mind makes to the interpretation of a single set of sensory data. They further reinforce the case by historical analysis of the ways in which different perceptions are allowed to "count" as relevant or irrelevant to a given scientific discipline at a given time.

Based on this range of information, the newer school of thought rejects the idea of a raw sensory "given" at the foundation of knowledge. Rather, it asserts that perception results from a combination of raw sensory "input" *and* the categories for its interpretation, as bestowed by socialization, education, or indoctrination, in a particular scientific discipline or culture. People are predisposed to "see" (that is, to recognize as significant or particular) entities or results that their predispositions condition them to expect to encounter or to find significant. As Stephen Toulmin has written, "What concepts a man employs, what standards of rational judgement he acknowledges, how he organizes his life and interprets his experience: all these things depend—it seems—not on the characteristics of a universal 'human nature,' or the intuitive self-evidence of his basic ideas alone, but also on when he happened to be born and where he happened to live."[8]

Referring to the development of scientific observation, Kuhn notes that an essential part of training in a scientific discipline is learning what standard observations and procedures count in that discipline as "seeing an entity" or "making an observation." A reading on a dial, for example, can be observed equally by a scientist in the course of an experiment or by a layperson, and they can even agree about the number presented by the dial. They differ, however, in what they observe. To the layperson, the number on the dial may be meaningless. To the observer familiar with the equipment and aware of the framework of the experiment in progress, however, it may be "seen as" an indication that, for example, a wire has begun superconduction or that equipment is malfunctioning.

It is crucial to recognize that the trained observer did not first note the dial reading, and then engage in a secondary process of deduction to determine meaning. Rather, once familiarity with such equipment and experiments is established, the observer simply *sees* what the reading represents. Kuhn writes, "What a man sees depends both upon what he looks at and also what his previous visual-conceptual experience has taught him to see."[9]

A person who trains in a particular science learns to see events and experiments in a particular way, to acquire a particular gestalt: "The world that the student then enters is not, however, fixed once and for all by the nature of the environment, on the one hand, and of science, on the other." Instead, the scientific world is determined by a complex interplay between discipline and experience; "it is determined jointly by the environment and the particular normal-scientific tradition that the student has been trained to pursue" (SSR, 111–12). Indeed, a researcher can perceive scientifically only after having acquired the relevant gestalt:

> Only after the researcher has learned to see reality in terms of accepted theory is research possible, but it is also possible for the researcher to discover anomalies and thus come to reconsider accepted theories. Two factors operate here. First, theories often provide a definite description of what the scientist ought to see and thus sharpen his vision for the discovery of anomalies. Second, as long as the scientist is carrying on empirical investigation it is not theory alone which determines what will actually occur, but theory in conjunction with a theory-independent world. Whenever the structure of theory and the structure of the physical world fail to mesh, anomalies will appear and although many anomalous events may eventually be interpreted in terms of accepted theory, it is the recalcitrant anomalies that eventually lead to the overthrow of one theory and its replacement by another, i.e., to scientific revolutions. (PTC, 108–09)

The theoretical framework the scientist brings to experiments or observations so affects perception that changes in the framework make it possible to "see" for the first time things that had, in another sense, been "seen" all along (SSR, 114–15). Indeed, often only after there is a place or concept for the observation in question within the dominant theoretical framework of the science of the time is the "observation" made at all. For example, Kuhn notes that Western astronomers, governed by a theory of a perfect and immutable heavenly realm, failed to see evidence of change in the heavens (such as sunspots or comets) until after Copernicus's theory had provided a context within which such changes were conceivable. In contrast, the Chinese, with even simpler equipment but without any belief in the immutability of the heavens, made such observations centuries earlier.[10]

Perception, then, is not a matter of being presented with "raw feels" or wholly uninterpreted sensory data on the order of the famous "red patch now" and then piling onto the foundation more abstract claims in a hierarchical fashion. Rather, an observer directly perceives what is produced by a combination of the sensory input and the categorial scheme that provides the context for such perception (*SSR*, 196).

This alternative theory of perception also implies a different theory of rationality than that which dominated the Logical Empiricists' approach. For them, rationality consists in the deduction by rigorous logical means of the implications and connections between propositions that were themselves grounded in indubitable ("incorrigible") observation statements. The newer view holds that there are no such theory- or context-neutral observations. Rationality can no longer be identified with deduction from indubitable perceptual foundations; rather, it must be analyzed with reference to the historical and sociological context of the reasoner (*HU*, 84).

The understanding of perception which dominates this analysis of the working of scientific knowledge is intimately linked with the direction of philosophical attention toward historical and sociological questions. The following concerns take on special importance: (1) the ways in which concepts and terms acquire meaning and are judged "true" in a particular discipline or theory; (2) the way in which disciplinary groups determine their boundaries, and educate, discipline, and acknowledge their members; (3) the analysis of the way in which knowledge within a given scientific discipline can be said to develop; and (4) the manner in which the background assumptions that govern a research enterprise are developed, transmitted, and transformed during the course of the historical evolution of a discipline.

THE NATURE OF SCIENTIFIC DISCIPLINES

A major point of departure for a new approach to the understanding of science is the observation that scientific research progresses and develops as it does in contrast to other kinds of inquiry and knowledge precisely because it focuses rather narrowly on the kinds of questions that are "scientific." While some diffuse inquiries may simply collect facts as they are encountered, without any particular organized method, the natural sciences are uniquely successful in generating "research questions" that are widely recognized within the discipline and can be universally recognized as crucial among the practitioners of the discipline.

The older philosophy of science accounted for this fact about scientific disciplines by claiming that the rigor of the link between observation and the theoretical claims determined progress. The new approach asks for real "explanations" and points out that explanations must be satisfactory with reference to the person or group in question: "explaining is a form of discourse which takes place in a specific context among some group of speakers and . . . instead of analyzing the explandandum and the explanans . . . we should, perhaps, examine the explainer and the explainee. The explainer's task is to get the explainee to understand something; anything that the former can say or do which will accomplish this purpose counts as an explanation" (*PTC*, 56).

This observation is commonplace in ordinary discourse. When a three-year-old child persists in asking "Why?" in response to every proffered explanation of a thing or event, adults express exasperation and may finally exclaim, "Just because it is!" Underlying this response, and underlying all explanation, is an implicit recognition that part of maturation into the world of common adult discourse is precisely learning when to stop asking "Why?" Of course, there is nothing sacred about where a society chooses to draw the line between intelligent questioning and frivolous or perverse persistence. Whereas in one society or subculture, the response "Because the Bible tells me so" may adequately justify a moral or scientific claim, it may not do so in another. Where one group accepts the assertion that a particular illness is caused by spirits or is punishment for a particular misdeed, another will not. But the explanations are accepted as fully adequate to the members of the society in the cultures or historical periods that offer and accept them.[11]

The new philosophers of science argue that scientific communities are similar to this larger adult-wide community of explanation. Scientific communities also socialize new members into their communities by teaching them the proper questions to ask. In addition, they teach them the "agreed results" of the discipline at a particular time, and those results provide touchstones on which future research projects are modeled and the results of those projects evaluated. Disciplines also train new members when to stop asking questions that would go beyond the range of legitimate inquiry as defined by the group at that point in its history.

Such limitations of inquiry and legitimate questioning are not inherent in the nature of things or in the inherent character of the subject matter of the discipline. Rather, they are a function of the sociological and historical character of the discipline at a specific point in its historical development.

Concepts must not be understood as free-standing points of definition, bearers of "meanings" in the abstract, but rather as terms that derive meaning by their location in a "constellation of concepts and the propositions and formulas in which they occur" (*HU*, 118–19). "A scientific concept is a knot in a web; the strands in the web are the propositions that make up a theory; the meaning of a concept is its location in the web" (*PTC*, 120).

Questions regarding a concept may pose a genuine research problem in one context and not in another. For example, the question "Why do heavy objects fall?" borders on being nonsensical in an Aristotelian framework, where "heavy" means, roughly, "objects whose natural place is down." "Heavy objects fall" turns out, when the meaning of "heavy" within this theoretical framework is explicated, to be nearly tautological. In a Newtonian framework, however, the answer, "Because gravity is acting upon it," leads to interesting lines of inquiry about the variability or constancy of the strength of gravity, the validity of the inverse square law, and so forth.

To extend the example, the question of the nature and source of gravity is itself a "non-question" in the Newtonian framework, where gravity simply "is." It becomes a question worthy of inquiry in research guided by the Einsteinian theory that gravity is not an irreducible concept but a feature of the structure of space-time.

In short, science does not simply read "off the face of nature" what questions need to be resolved or investigated. Rather, questions become justified based on the location of the terms and concepts they comprise, within the overarching framework from which they get their meaning.

Gravity, an irreducible simple in Newton's scheme, is not so in Einsteinian physics. Rather, it becomes there a complex notion subject to further analysis and description.

A particular research community, then, is a community that broadly agrees on the meaning of the concepts it uses, the outstanding issues for resolution within that discipline, and the techniques and equipment that are appropriately used in the effort to resolve questions. Only when there is broad agreement within a group regarding such matters can the most typical activity of the scientist—what Kuhn calls "normal science"—begin.

Although Kuhn's use of the term "paradigm" is notoriously ambiguous,[12] it is precisely this general feature of intellectual disciplines which he seeks to point out. What he calls "normal science" is research devoted not to questioning or challenging the fundamental assumptions shared within the discipline at a particular time, but rather to "puzzle-solving" within the context of those assumptions: "one of the things a scientific community acquires with a paradigm is a criterion for choosing problems that, while the paradigm is taken for granted, can be assumed to have solutions" (*SSR*, 37).

Normal science is concerned with three basic kinds of activities. First, it attempts systematically to determine facts that the theory suggests are useful or significant, with precision and/or in a broader range of cases than those in which they are already known (for example, stellar position and magnitude for astronomy). Second, it seeks to determine facts that can be directly compared with the predictions of the paradigm theory. Third, it does empirical work "undertaken to articulate the paradigm theory . . . resolving some of its residual ambiguities and permitting the solution of problems," such as the determination of the numerical value of physical constants.[13]

For normal science—the activity that engages the vast majority of scientists for the greater part of their time—progress is impossible until the essential theoretical structure that should be brought to bear on the empirical world has been accepted. As Kuhn notes, "Normally, the members of a mature scientific community work from a single paradigm or from a closely related set" (*SSR*, 160–61).

Those disciplines we unhesitatingly call sciences are those that have settled into the pattern referred to by Kuhn as that of a "mature science." How does that settling in occur? Unlike the Logical Empiricists who attempted to dictate a priori the criteria a discipline must meet to be scientific, a field becomes scientific through a consensus not on the definition of science, but rather on the significance of accomplishments in a substantive field (*SSR*, 162).

Such consensus or agreement emerges only gradually from a prior "pre-paradigm" stage characterized by uncertainty and debate over the proper goals and methods of the emerging discipline. Kuhn's discussion of this process has been sharpened and improved by Margaret Masterman, who replaces Kuhn's simple distinction between pre- and post-paradigm periods with more subtle divisions: non-paradigm, multiple-paradigm, and dual-paradigm science.[14]

In the non-paradigm period, facts are collected, but without any guiding principle of research or organization. The book, not the article, is the major vehicle of publication. There are a number of competing schools attacking one another's fundamental assumptions and methods. As Masterman writes, "Non-paradigm science is barely distinguishable, if at all, from 'the philosophy of' the relevant subject."[15]

In sharp contrast are sciences with multiple paradigms, such as contemporary psychological and social sciences. Information is collected carefully and with guiding assumptions about what facts are relevant and important. But multiple sets of paradigms exist and correspondingly, competing schools, which agree about the subject matter they are concerned with, disagree on the appropriate framework within which to conduct their research. In this case, no unified discipline engages in research guided by a shared paradigm. Rather, the field is characterized by multiple subfields, and each of these continues to criticize or ignore the governing assumptions of competing subfields.

Progress and a sense of collective enterprise in the form of puzzle-solving characteristic of Kuhnian normal science is possible in such multiple-paradigm fields and periods, but it is "local" progress within one "school" rather than long-term progress that is recognized throughout the field in question.[16] Such a period is brought to an end in a field, if it is brought to an end at all, only when "someone invents a deeper, though cruder . . . paradigm, which gives a more central insight into the nature of the field, though restricting it and making research into it more rigid, esoteric, precise."[17]

Distinguishing these two kinds of periods, Masterman argues, allows clarification of Kuhn's notion of intellectual crisis that gives rise to scientific revolution. Periods of scientific crisis in a discipline are neither non-paradigm nor multiple-paradigm periods, but dual-paradigm episodes, during which there are two competing paradigms within the discipline, each of which has something to be said for it. Unlike the differences between schools in multiple-paradigm science, in dual-paradigm periods the differences between schools are much less global and philosophical. Rather, the differences center on the adequacy of the two respective paradigms in solving or guiding research toward solving what both groups agree are the outstanding explanatory problems of the field.[18]

What Kuhn calls normal science, therefore, can occur in a number of settings other than that of a mature science. Paradigm-guided puzzle-solving is also a feature of the research of a subfield during multiple-paradigm periods, as well as of each of the competing disciplinary groups during dual-paradigm episodes.

Masterman also notes that normal science can set in prematurely, when the paradigm that comes to be adopted is not, in fact, a useful guide to research. As an example she cites astrology, which is in many respects remarkably like normal science in that the practitioners agree on technique and procedure.[19] But she argues persuasively that the resemblance of such premature paradigm-guided "disciplines" coincides with our ordinary language for the description of such groups: we do not call them non-sciences, but rather pseudo-sciences.[20]

The point of this diversity of contexts for normal science is, therefore, that puzzle-solving within a shared framework of assumptions that define the relevant questions and issues for disciplined research can begin in a wider number of contexts than Kuhn's original framework suggested. Wherever it does occur, however, it is because some group (larger or smaller) has at least provisionally accepted both a model for the kinds of research that will be interesting and some framework within which that research can proceed.

How is it, then, that such communities of inquiry come to be formed and sustained? The philosophers of science we are considering typically approach this question with a sociological analysis of the process of educating and indoctrinating new researchers into the discipline (rather than on the attempt to formulate some a priori logic of discovery as the Logical Empiricists had done). What provides the unity to such groups is what Toulmin calls their shared intellectual ideals (*HU*, 151). He means that a group shares a common conception of what "general forms . . . a complete account of [the area of inquiry of the discipline in question] should take" (*HU*, 151). The presence of these shared ideals generates the questions that the discipline is supposed to answer, and provides the terms in which those issues and expected answers are cast: "Scientists locate and specify the intellectual shortcomings of their current concepts by recognizing the shortfall between their current ability to 'account for' the relevant features of the natural world and the explanatory ambitions defined by their current ideals of natural order, or models of complete intelligibility" (*HU*, 152).

To train in a particular scientific discipline, then, requires an apprenticeship during which one learns to enter into the intellectual ideals of the discipline in question. This is the prerequisite of scientific research, since it is "only by matching natural phenomena against the intellectual template of these ideals" that one can identify the "characteristic problems" of a particular discipline (*HU*, 153). In other words, a function of the process of scientific education is to precondition the student to think and work within the shared framework of the group.

The effect of this preconditioning of a group engaged in disciplined research is a degree of incommensurability or even inability to communicate among individuals engaged in research governed by differing assumptions. This is true both in the case of contemporaries bound by differing models and of individuals from different historical periods in the evolution of the same discipline,[21] and is another way of stating the observation made earlier regarding the theory-laden character of perception. About disputes between individuals guided by different paradigms Brown observes, "One of the most striking characteristics of debates between thinkers who are working from different presuppositional bases is that along with disagreements on what problems need to be solved and what constitute adequate solutions to these problems, they also disagree on which concepts do and which do not require explication and find it unnecessary (if not impossible) to offer arguments for their choice" (*PTC*, 57). A major source of this difficulty in communication is that while individuals governed by different paradigms may be using the same word, they often are using it to refer to a concept with different meaning because of its different location or function within the larger theoretical structure defined by the paradigm.

For example, the terms "weight" and "fall" function in both Aristotelian and Newtonian mechanics, but refer to concepts that are neither entirely dissimilar nor identical. "Fall" for the Aristotelian is "motion to natural place"; for the Newtonian it is "motion under the influence of gravity" (*PTC*, 116–17).

One consequence of differences of this sort is that propositions that are axiomatic in one system may be entirely nonsensical in another. For example, the modern astronomical commonplace, "Our sun is a star," would have been literally nonsense to a medieval, for whom the concept "sun" (one of the planets) was by definition distinct from the concept "star" (one of the fixed objects in the celestial sphere) (*PTC*, 116–17).

Since only the emergence of a paradigm makes normal science possible, it is important to ask how such a paradigm does emerge, and the kinds of reasons that a group may have for adopting it at the time of its emergence. If only the emergence of the paradigm permits careful distinctions between the "rational" and the "nonrational" within the context of that discipline, can the acceptance of the paradigm in the first place itself be a "rational" thing to do? Must it necessarily be nonrational, if not irrational? And if it is the latter, is the implication that the very model of modern rationality, the scientific method, is itself grounded on the nonrational?

The starting point for discussing this question in this tradition of the philosophy of science centers in the analysis of the behavior of scientists and disciplines during periods of scientific crisis in which one dominant model for guiding research in a discipline comes to be replaced with another. Indeed, much of the debate is concerned with the frequency, extensiveness, and rationality of such transitions.

The most famous position, as well as one of the most extreme, is that taken by Kuhn in the first edition of his *Structure of Scientific Revolutions*. There, he draws a sharp contrast between normal scientific activity, which is preoccupied with problem-solving in a very narrow framework of a shared paradigm, and periods of extreme theoretical crisis during which the previous paradigm is generally known to be inadequate for the solution of the issues that the discipline itself recognizes to be most pressing. He calls these kinds of crises scientific revolutions, and he characterizes them as periods in which there is a shift from the previous paradigm to a new one, which progressively redefines the discipline itself and sets research off in a new direction.

On this analysis, two questions are of greatest interest: How does a new paradigm come to be accepted? And to what extent is there continuity within a disciplinary group across the boundaries of such shifts? Kuhn's position is the one that has evoked qualification and response from other members of this tradition.

On the first point, Kuhn is concerned, especially in the earlier formulation of his position, to stress the nonrational factors that account for the acceptance of a new paradigm. Citing the acceptance of Copernicus's theory and the supplanting of Ptolemaic astronomy, Kuhn writes, "Copernicus' more elaborate proposal was neither simpler nor more accurate than Ptolemy's system. Available observational tests . . . provided no basis for a choice between them" (*SSR*, 75–76). In fact, Kuhn argues, the only reason that Copernicus's theory was accepted was the feeling that the Ptolemaic system had simply failed to solve the problems it had set for itself for such an extended period of time that "the time had come to give a competitor a chance."[22] In summary, Kuhn's position holds that "Paradigm change cannot be justified by proof" (*SSR*, 152).

Kuhn criticizes Popper's view that theoretical frameworks can be decisively falsified, thereby providing a means for determining the better of competing paradigms. Rather, Kuhn argues strenuously that the disagreement of different groups on the scientific problems they must address and the standards with reference to which the problems are to be assessed makes the process of the determination of a better paradigm impossible to schematize:

> If there were but one set of scientific problems, one world within which to work on them, and one set of standards for their solution, paradigm competition might be settled more or less routinely by some process like counting the number of problems solved by each. . . . [But] the proponents of competing paradigms are always at least slightly at cross-purposes. Neither side will grant all the non-empirical assumptions that the other needs to make its case. (*SSR*, 147–48)

Kuhn stresses the ways in which, rather than having a clear and formalizable logic by which to change paradigms, acceptance of paradigms is conditioned, if not determined, by nonrational and idiosyncratic circumstances of researchers: "Individual scientists embrace a new paradigm for all sorts of reasons and usually for several at once. Some of these—for example, the sun worship that helped make Kepler a Copernican—lie outside the apparent sphere of science entirely."[23] In short, "as in political revolutions, so in paradigm choice—there is no standard higher than the assent of the relevant community (*SSR*, 94).

On the question of the continuity between different historical periods governed by differing paradigms within the same discipline, Kuhn is equally forceful in his position: "After a revolution scientists are responding to a different world" (*SSR*, 111).

Stating the position even more strongly, Kuhn writes:

> In a sense that I am unable to explicate further, the proponents of competing paradigms practice their trades in different worlds. . . . Practicing in different worlds, the two groups of scientists see different things when they look from the same point in the same direction. . . . In some areas they see different things, and they see them in different relations one to the other. That is why a law that cannot even be demonstrated to one group of scientists may occasionally seem intuitively obvious to another. Equally, it is why, before they can hope to communicate fully, one group or the other must experience a conversion that we have been calling a paradigm shift. Just because it is a transition between incommensurables, the transition between competing paradigms cannot be made a step at a time, forced by logic and neutral experience. Like the gestalt switch, it must occur all at once (though not necessarily in an instant) or not at all. (*SSR*, 150)

To exemplify the subtle continuities and discontinuities within a discipline that has undergone a shift or change of its underlying assumptions, Brown discusses the frequent claim that the theory of relativity is merely a generalization of Newtonian mechanics, which reduces to Newtonian description in low-velocity situations (low velocity with reference to the speed of light, of course).[24]

Brown agrees that the equations derived from the theory of relativity are formally identical to those derivable from Newtonian mechanics in such situations. But he argues persuasively that formal identity alone is an inadequate criterion for "sameness," since the question remains whether the concepts involved are identical in their reference and in their connections with other concepts within the framework of the theory. When the latter questions are addressed, it can be shown that the concepts involved in the Newtonian framework are significantly altered in their transplantation into the relativistic framework, even if the equations yield identical results in some contexts, and even if the term used to label them remains the same. For example, "for Newtonian mechanics the mass of a body is constant; for relativity theory the mass of a body is a variable dependent on velocity, with a minimum value equal to the mass as given in Newtonian theory and an upper value which increases without limit as the relative velocity of a body approaches the velocity of light."[25]

On the other hand, in some ways concepts remain the same in both contexts. For instance, mass is understood in both frameworks as a measure of the resistance to change in velocity. But "the way in which velocity changes has been reformulated, as well as playing a role in the concepts of energy and momentum" (*PTC*, 123).

The situation is similar as to the concept of time in the two contexts. Time is an absolute and universal property in Newton's framework; it is a function of relative velocity between reference frames in the relativistic context. So, while "time" still does "many of the same jobs" in the newer framework, it is not identical to the older concept. But neither is it absolutely discontinuous with it. Rather than being a generalization of the older concept, it is a "modification and development" of it (*PTC*, 123).

Although identical formulas may be derivable from the two theories, the concepts and terms used in those formulas are not identical. "Understood in the context of their respective theories, they say different things about the structure of the physical world" (*PTC*, 123).

The relationship between successive paradigms within a discipline is not, therefore, usefully analyzed as either one of identity or as one of radical discontinuity. In many instances it is rather a far more subtle realignment of a "web" of concepts and the threads that interconnect the concepts representing the nodes of that web. Since a theory that comes to be supplanted by a new and richer one—if it had any significant period of existence—presumably succeeded in describing or predicting results at least in some of the same areas of experience, it is not surprising that the new theory should be able to yield identical results in some cases. But the process of realignment is subtle, and it requires close analysis of particular episodes of such shifts to determine exactly how much changes and how much remains the same in each case.

THE PROBLEM OF RELATIVISM

The most troubling and difficult issue raised by the new analysis of the science is that of relativism. If the means by which a new paradigm comes to be accepted in a particular discipline is the result of many factors, some of which are inherently nonrational, and if theory conditions not only the meaning of empirical evidence but even the degree to which a particular kind of evidence will be thought to be significant, or even present, then how are we to account for the intuitive sense we have that the sciences tell us about the world as it is? What place remains for the claim that the sciences, or any forms of human knowledge, yield "truth"?

For the philosophy of science of the Logical Empiricist period the approach to answering the question of relativism is clear: Scientific theories are built up from indubitable foundations of empirical observation, directly grounded in the world as it is. Therefore, as scientific theories are articulated, they approach ever more closely the explanatory goal of providing a mirror-image of the connections between things as they were objectively in the world. As Richard Rorty characterizes this presupposition, "Scientific inquiry is supposed to discover what sorts of objects there are in the world and what properties they have. Anybody who conducted serious inquiry could only have been asking which predicates were to be pinned on which things."[26]

But having rejected the notion of founding knowledge on such incorrigible empirical data, the newer philosophical tradition cannot make use of this general line of explanation. How, then, can it account for the fact that scientific theories within a discipline "progress" without claiming that the measure of their progress is an increasing approximation to the way things objectively really are?

Rather than imagining theories to achieve ever-greater approximation to the structure of the world in and of itself, the general analogy

24

preferred by Kuhn to illuminate scientific change is one drawn from evolutionary biology.[27] In biology it is necessary to account for organisms' increasing complexity and greater adaptation to a range of environments, but it is incorrect to imagine that process as developing toward a predetermined telos.

Analogously, in the development of the sciences it is possible to treat scientific development and progress by analyzing its movement from primitive beginnings without imagining that there is a goal given in nature itself toward which it evolves. Kuhn summarizes the analogy thus:

> The process described . . . as the resolution of revolutions is the selection by conflict within the scientific community of the fittest way to practice future science. The net result of a sequence of such revolutionary selections, separated by periods of normal research, is the wonderfully adapted set of instruments we call modern scientific knowledge. Successive stages in that developmental process are marked by an increase in articulation and specialization. And the entire process may have occurred, as we now suppose biological evolution did, without benefit of a set goal, a permanent fixed scientific truth, of which each stage in the development of scientific knowledge is a better exemplar. (*SSR*, 172–75).

This process of progressive improvement in scientific theory can occur because, even though it is true that perception is theory-laden and conditioned, it is important not to confuse that claim with a fully idealist position that perception is reality, pure and simple. However strongly the claim is made that theory conditions perception, the claim cannot be that theories create their own data. Rather, "the objects of perception are the results of contributions from both our theories and the action of the external world on our sense organs. Because of this dual source of our percepts, objects can be seen in many different ways, but it does not follow that a given object can be seen in any way at all" (*PTC*, 93).

The point of drawing this distinction is that, while it may be that what one perceives is significantly affected by the theoretical framework that one brings to experience, there are external boundaries beyond which legitimate interpretation does not go. Using the duck/ rabbit gestalt diagram as an example, Brown writes, "This figure can be seen as a duck, a rabbit, a set of lines, or an area, and one might plausibly imagine its being seen as a piece of laboratory apparatus, a religious symbol, or some other animal by an observer with the appropriate experience. But try as I will, I cannot see this figure as my wife, the Washington monument, or a herd of swine" (*PTC*, 93).

The supposed dichotomy between perception as purely passive receipt of objects that simply are what they appear to be and imagining perception as absolute creation of perceptual objects is a false one. Instead, "we shape our percepts out of an already structured but still malleable material" that "limit[s] the class of possible constructs without dictating a unique percept" (*PTC*, 93).

Furthermore, although all perception may be preconditioned by the assumptions that the perceiver brings to bear on them, it by no means follows that there are no distinctions to be drawn regarding the significance and value of differing perceptions.

Brown illustrates the point, citing the example of two people passing a steel mill. One may only recognize the smell of rotting eggs, or perhaps only that the smell is unpleasant, while the other may be able to designate it precisely as the smell of sulfur dioxide. While both perceptions are correct and both arise appropriately from the experience in question, they are not on an epistemic par. Rather, the second perception has "greater epistemic value" and gives rise to more knowledge because its labeling links it with a vast structure of chemical theory which allows the observer to infer information regarding the processes by which the smell came to be present, the implications of the odor regarding the gravity of pollution being generated, the connection between the odor and acid rain downwind, and so forth.[28]

Hence, the thrust of the epistemology developed in order to describe the actual working of science is to find a "middle way" between the model of strict deduction from indubitable and uninterpreted sensory data that characterized the Logical Empiricist approach and a full-blown idealism that denied any connection between the literal offerings of sense and the interpretations that can be placed upon it. It attempts to show that a simple dichotomy between relativism and absolutism is unable to account for the actual epistemological situation, and that a significant degree of relativity of perception must be acknowledged if one is to account for the facts of the historical development of the sciences. On the other hand, as the discussion has matured from the earliest of Kuhn's dramatic pronouncements, it has become clear that an absolute idealism (with correlative absolute relativism) is equally unable to account for the whole of scientific rationality. Instead of this too simple polarity, the emphasis is increasingly on progressive and evolutionary improvement in the fit between theoretical explanation and the data of sense. This model denies the realism of the goal that theoretical accounts strive to be literal mirrorings of the external world, while preserving a significant sense in which successor theories are more adequate for the explanation of the world than their predecessors.

Central elements in the epistemology that has emerged in the analysis of the working of the natural sciences need to be noted. First, each emphasizes the contribution of the observer to the observation. Second, each accepts a degree of relativism into the core of epistemology that distinguishes its positions from those of its predecessors. Finally, each claims that substantive intellectual construction begins, whether in theology or in a particular scientific discipline, only when shared assumptions and frameworks within which that inquiry and conceptual elaboration are to proceed have been agreed upon by the relevant community of research.

2 | THE SPECTRUM OF CONFESSIONAL THEOLOGIES

T HE CENTRAL CLAIM OF THE SPEC-
trum of confessional theological methods is
that theology must proceed from essentially inner resources of the
Christian community and tradition rather than basing or grounding
itself in commonly available cultural convictions. Confessional thinkers
disagree over the details, but they all share a concern to maintain the
integrity of the Christian tradition and community over against the
corrosive forces of the intellectual fashions and concerns of the age.

Opponents of confessional method charge that this concern necessarily leads in a fideist or internalist direction. For them, confessional
method seems to assert that theology is immune to criticism and
revision because such criticism lies outside the inherited theological
tradition or the scope of revelation (however that is understood). This
is a distortion of the thrust of confessional theology which results in
objections to confessional theology in general that although applicable
to many inhabitants of the district, fail seriously to engage others. In
particular, H. Richard Niebuhr's understanding of resolute confessional
method is distinctive among basically confessional methods in that it
is generally not subject to the more damaging criticisms leveled at
confessional method.

To correct the distortion, alternative understandings of confessional
method need to be established. Justification and implications of con-

fessional method as advocated by representative confessional theologians must be clarified. The work of such theologians as Karl Barth, who is the seminal figure in neoorthodox confessionalism; Paul Holmer and D. Z. Phillips, representatives of a theological tradition based on the later philosophical work of Ludwig Wittgenstein; George Lindbeck, and Francis Schüssler Fiorenza, a Roman Catholic, will serve, when seen together, to achieve these goals. In the course of this analysis, the features that unify them will be identified. In addition, the legitimacy of the most damaging charges made against some versions of confessional theology will be examined, and the criteria that an adequate confessional theology needs to meet will be noted.

KARL BARTH'S PROPOSAL FOR KNOWLEDGE

Barth views theological language as ordinary language operating in the peculiar environment of given revelation. Barth's thought is founded upon substantive theological claims. In these respects, Barth differs significantly from Holmer and Phillips, who attempt to show how religious discourse differs from other modes of speech and to use that philosophical claim as a basis for confessional method.

But like all confessional thinkers, Barth rejects any understanding of theology that appears to be founded in the claims of other intellectual disciplines. He writes,

> Talk about God has true content when it conforms to the being of the Church, i.e., when it conforms to Jesus Christ. . . . It is in terms of such conformity that dogmatics investigates Christian utterance. Hence it does not have to begin by finding or inventing the standard by which it measures. It sees and recognizes that this is given with the Church. It is given in its own peculiar way, as Jesus Christ is given, as God in His revelation gives Himself to faith. But it is given. It is complete in itself. It stands by its claim without discussion. It has the certainty which a true standard or criterion must have to be the means of serious measurement.[1]

Indeed, dogmatics "knows the light which is intrinsically perfect and reveals everything in a flash" (*CD*, 1.1:14), and therefore must not cast about for a "foundation," as if uncertain of its point of departure.

Barth's reason for such strong assertions is theological rather than epistemological, at least in the first instance. Theology is critical inquiry into the church's distinctive speech about God, but it must be critical with reference to a standard of criticism that, in fact, corresponds to the distinctive existence of the church. The only foundation for that distinctive mode of existence, Barth argues strenuously, is "Jesus Christ, God in his gracious revealing and reconciling address to man" (*CD*, 1.1:4). All speech that takes its point of departure elsewhere, therefore, is doomed from the outset to be inadequate to the divine subject matter of theology. Indeed, for Barth, the major mistake of Liberal theology lies in its loss of its "proper criterion" of critical assessment (revelation) in favor of a desire to render the church and its message meaningful in the modern context (*CD*, 1.1:251ff.). How, Barth asks, can another criterion be substituted and the path not be lost? (*CD*, 1.1:252).

Barth discusses the possibility of a foundation to theological reflection at greatest length with reference to the question of the possibility and nature of "prolegomena" to dogmatics. Barth does countenance a particular kind of prolegomenon, that of combating the "heresies" of Catholicism and Liberalism (*CD*, 1.1:31). But he strenuously rejects the kinds of apologetic offered for Christian faith since the Enlightenment: appeals to general canons of reason and the knowledge provided by other disciplines of inquiry, and attempts to build upon them to show the rationality of belief or the superiority of Christianity to other religions. Such apologetics are misguided and even dangerous, Barth argues, for three reasons. First, by seriously treating unbelief, dogmatics betrays an uncertainty about itself and its nature which it ought not to have. Second, it seems to reflect a belief (a false one, according to Barth) that dogmatics itself is clear regarding its content and has accomplished its task, leaving only the necessity of persuading the outsider to become interested in its conclusion. And last, beginning with apologetics is an irremediable misstep at the outset because it operates as if, once the apologetic task is accomplished, dogmatics can safely go about its business without the continuing exposure of its methods and conclusions to the critical questions raised by outside concerns.[2] He summarizes his conclusion thus:

> In order to give an account of the way of knowledge pursued in dogmatics, we cannot take up a position which is somewhere apart

from this way or above the work of dogmatics. Such a place apart or above could only be an ontology or anthropology as the basic science of the human possibilities among which consideration is somewhere given to that of faith and the Church. . . . Evangelical dogmatics cannot proceed along these lines. . . . It realises that all its knowledge, even its knowledge of the correctness of its knowledge, can only be an event, and cannot therefore be guaranteed as correct knowledge from any place apart from or above this event.[3]

Rather than trying to show the parallels and congruences between theology and other aspects of knowledge and culture, "scripture should . . . be and become and remain the master in theology's house" (*CD*, 1.1:285). The theological reason for this restriction on the foundation of theology is the following:

We are dealing with the concept of revelation of the God who according to Scripture and proclamation is the Father of Jesus Christ, is Jesus Christ Himself, and is the Spirit of this Father and this Son. Naturally one might also investigate the concept of completely different revelations and perhaps even a general concept of revelation. But in so doing one would abandon the task of dogmatics. For it is this concept and this concept alone, that interests dogmatics. From this concept and this concept alone can it expect light on the relation between the Bible and Church proclamation and guidance in criticism and correction of the latter. If we are not to make a transition to some other discipline which may well be good and necessary but is not dogmatics, then at this point where the relation between the Bible and Church proclamation is in question we can only investigate this concrete revelation. (*CD*, 1.1:291)

There is a clear difference between grounding the independence of theology in epistemological or in theological principles. Barth's treatment of the question whether that independence is strictly necessary or only *de facto* required under particular circumstances brings that difference to light. Holmer and Phillips (as we will see), for example, consider the strict independence of the religious language game with reference to others a permanent and logically necessary feature of (true) religious discourse. In contrast, Barth writes, "Even the asserted independence of theology in relation to other sciences cannot be proved to be necessary in principle. . . . Theology does not possess special keys to special doors. Nor does it control a basis of knowledge which might not find actualisation in other sciences. . . . Philosophy and secular science generally do not have to be secular or pagan. There might be such a thing as *philosophia christiana*" (*CD*, 1.1:5).

Rather than being a permanent and necessary feature of religious language, the existence of theology as separate in method and content is, for Barth, an "emergency measure" that has "no epistemological basis" (*CD*, 1.1:7). Indeed, "from the standpoint of the Church itself and therefore of theology, the separate existence of [the other sciences] is theoretically very questionable" (*CD*, 1.1:7). In fact, "all sciences might ultimately be theology" (*CD*, 1.1:7), but are not because of their refusal to adopt Jesus Christ as the fundamental principle of criticism for their disciplines as well.[4] The existence of the pluralistic and un-integrated world of sciences (academic disciplines) is, for Barth, evidence of the effects of the Fall. Were humanity not separated from its true ground and standard in God, it would recognize that all true knowledge acknowledges that source in God.

Furthermore, having said that the independence of theology is a necessity under the present circumstances, Barth guards against a notion that theology possesses a "divine magisterium," one that is the same from age to age,[5] in order to guard the divine transcendence against the presumption of human claims to knowledge. Against any such claim to certainty or constancy of dogmatic knowledge, Barth asserts:

It is by no means the case that in dogmatics the Church becomes as it were the lord and judge of the subject-matter, so that the current results of dogmatics are to be accepted as a law imposed as it were on God, revelation and faith. Dogmatics has to investigate and say at each given point how we may best speak of God, revelation and faith to the extent that human talk about these things is to count as Church proclamation. It should not think that it can lay down what God, revelation, and faith are in themselves. In both its investigations and its conclusions it must keep in view that God is in heaven and it on earth, and that God, His revelation and faith always live their own free life over against all human talk, including that of the best dogmatics. (*CD*, 1.1:85–86)

In summary, then, Barth's major concern is to avoid foundationalism, the belief that dogmatics can begin by showing the coherence of its claims and methods with those of philosophy, anthropology, or other nontheological disciplines. This concern places him squarely in the center of the impulses that drive and inspire confessional method in theology.

Unlike the Wittgensteinians, Barth rejects theorizing about the uniqueness *in principle* of religious language. He insists instead on the theological point that dogmatics must be governed by and assessed with reference to criteria that are unique to its distinctive mode of speech. This means that the Word of God alone must govern its speech, and since that word has an "event" character, it can never be treated as one element of a larger theory. That would make it captive to a human theoretical framework.

In contrast to Niebuhr, Barth's position also rejects the attempt to explain in some more general philosophical and social scientific way how the distinctive language and symbols of the inner history of the Christian community come to be transmitted and maintained within the community. The refusal has a justification and an integrity, given Barth's view of revelation. On its own terms, it is neither internalist nor fideist. Whereas all sciences in principle could be integrated into a *philosophia christiana,* the present internalist stance, according to Barth, is provisional and temporary. And it is not fideist because the alternative—starting outside the assumptions of revelation—is not conducive to doing theology in the proper sense. True knowledge of any kind is only possible, on Barth's terms, from inside, since all other attempts are a snare and a delusion, a mere projection of human wants and wishes.

Unless one is prepared to step fully into the thought-world of those assumptions, however, Barth's position does appear to be fideist and internalist. He holds out the possibility in principle of an integrated thought-world that would include theology and all other disciplines, when in fact there is an unbridgeable gulf between them under the conditions of finitude and sin.

The overall judgment to be made about the Barthian position is that it rejects at every turn the possibility (not to mention the necessity) of general theoretical formulations regarding methodology in favor of language that is dynamic, that speaks of the Word of God as "making itself known." From Barth's point of view, the greatest need in dogmatics is to preserve this notion of God "speaking," of the freedom of God's Word to address us when and how God chooses to do so. That this has happened and continues to happen is, according to Barth, the axiom from which all dogmatics must proceed.

Barth's position is, at least in its own terms, insulated from philosophical analysis; it explicitly rejects the subordination of theological/ dogmatic utterance to that or any other human criterion. This may, from a certain theological perspective, be to its credit, but it requires a suspension of philosophical standards of analysis which ordinarily apply to any realm of discourse, and the acceptance of powerful theological affirmations on what must, from outside the Barthian framework, necessarily seem to be fideist grounds. Barth's fundamental concern to preserve the integrity of distinctively Christian affirmations can be met without the need to isolate theological reflection so absolutely from other intellectual disciplines.

In contrast to Barth, Paul Holmer and D. Z. Phillips insist that they are subject to the criteria of philosophical assessment, and reject only what they consider to be the importation of inappropriate philosophical criteria. Those false criteria, they claim, are not derived from the close analysis of the way in which religious language works, but from an a priori philosophical framework arbitrarily asserted as universal across the boundaries of language-games.

PAUL HOLMER'S CONFESSIONAL THEOLOGY

Paul Holmer's theological work poses some difficulties. He is crystal clear in describing what he is against: the separation of academic theology from the religious life, the failure of theological writing to include the invitation for personal transformation, and so forth. His efforts to sketch a coherent philosophical and theological position that can support those concerns, however, are less satisfactory. Among the theologians treated herein, perhaps Holmer comes closest to warranting the suspicion that confessional method is a code-word for the defense of an unyielding orthodoxy or for pietism.[6] But he has made other assertions that are not obviously integrated into the structure of his thought, but at least counter such an interpretation.

Holmer's central concern is that theology has become detached from the real business of religious living. He writes that theology "is divorced from the plain elements and words that fashion the Christian's faith. Certainly it is time to get theology hooked up once more with the clean breeze of God's grace that blows through the centuries. The outlook of this age or that is swept aside, as room for the Gospel is once more established. Theology states those simple 'credenda,' those rudimentary themes that make every age a part of God's great symphony."[7]

The cause of this separation, as Holmer diagnoses it, is that theology has been bewitched by other disciplines (such as history and the sciences) into thinking that it must operate with language in the "about" mode in order to be rigorous or, more strongly, even to have clear meaning.[8] But, he argues, theological language that is truly concerned to communicate about God as God is encountered in the actual religious life, does not encompass the impersonal "about mode"; rather, it is the very language of religion, as used by religious people, that is the proper language for theological expression.[9] Holmer asserts this but he grants that theology like Luther's or Paul's "almost seems by definition to be radically expressive, personalistic, confessional, illogical, and disorderly. . . . The 'confessional' category itself makes such reflection seem immediately nonobjective, nonlogical, and educationally irregular. Admittedly, theology of this . . . kind is everywhere permeated by an overpowering religious passion. Such language is not neutral and dispassionate" (SLLR, 63).

But, while not "neutral and dispassionate," Holmer nevertheless argues that the kind of theology he advocates is "not incorrigible or fantastic," but can be "thoughtful and logical without ceasing to be an accurate and forthright expression of a conscientious person in the grip of a religious passion" (SLLR, 63).

But if theology is not to be incorrigible, with reference to what is it corrected? Here, things get murky. Holmer asserts that "theology is not simply expressive . . . for it is also ordered by distinctively religious categories" (SLLR, 65), but unfortunately does not go on to specify what those categories are, either in that discussion or any other. And whereas in other contexts he seems to suggest that theology is incorrigible, one is left in confusion as to Holmer's true position. For example, he writes, "This is how theology finally realizes itself in its correct form. The teachings do not have to change at all, for they are kind of constant stretching through the ages" (WTID, 29). And again,

"I wish to insist . . . that the core of theology for Protestants, not only for Catholics, is a divine 'magisterium,' which is the same from age to age. . . . Theology proposes something that is timeless and eternal" (WTID, 12).

In fact, the notion that theology must change to be intelligible for new historical periods is one that Holmer explicitly rules out. "One must never entertain . . . a picture of a Christian theology as a net of causes and reasons, an intellectual proposal, which by constant assimilation of novelties, by continual adaptation to new circumstances, will reclaim the masses by its sweet reasonableness."[10]

If theology is concerned with a "magisterium" that is "timeless" and "eternal," how are we to account for the obvious historical fact that the theology we know is as prone as any other kind of inquiry to intellectual change and alteration in various environments? For example, religious explanation has retreated from areas now considered purely questions of biology or astronomy. How can we account for that retreat? Holmer argues that withdrawal of theology from such areas is probably desirable from the religious point of view, and does not really represent a loss of explanatory ground on the part of theology. Rather, it simply eliminates a confusion of intellectual spheres that should never have been present at all.[11]

The root problem for Holmer is the assumption of a "generic theory of meaning" which leads us to think that judgments can be made between various spheres of discourse (historical, theological, scientific, etc.) (SLLR, 68). Drawing on the philosophy of Wittgenstein, Holmer enthusiastically endorses the view that there is no single notion of "fact" that can be used between spheres of discourse, and that there are only various language games, each of which has a unique grammar and whose logic can be discerned only by the players of that game (SLLR, 65). There is no "subgame basic to all the rest" which can adjudicate disputes between various language games and determine which one really has a legitimate claim to coherence with the "facts" (WTID, 6).

Once the root problem is resolved, Holmer asserts, it becomes evident that there is no conflict between theological and scientific or historical facts. Each properly refers to its own kind of facts: theological facts for the theological game, historical facts for the historical game, and so on (WTID, 8). Because these different games and their correlative kinds of facts are "incommensurable" (that is, not subject to any common measure), incompatibility is a logical impossibility (SLLR, 73).

And so the world is made safe for the meaningfulness and ongoing vitality of Christian theology; it is only necessary to speak the "language of" Christianity. The possibility that cherished theological/religious beliefs may simply turn out not to be sustainable in the face of changing historical or scientific understanding has been eliminated not just for present scientific and historical understanding, but for all time. This has been achieved principally through the distinctive use of the word "fact"; for that reason, Holmer's discussion of terminology bears closer examination. "Fact," Holmer asserts, is a word we use to designate that which is not disputed within a particular language game. As he expresses it, "What is a 'fact' is so in this context or that. Usually what we call a fact is what we can reason from, what we can take for granted, what is agreed upon, and what (in this context, place, and for given purposes) we do not need to quarrel about and perhaps cannot" (MT, 105).

Holmer's claim is misleading because of its extremely unqualified nature. There are many communities of human beings that take various things for granted, and Holmer's position seems to grant them all equal status. "Astrological facts" or "facts of Scientology" acquire an intellectual status on a par with "historical facts" or the "facts of physics," simply because they are bodies of assertions about which (presumably) astrologers or Scientologists agree. If these various kinds of facts are granted presumptive equality in addition to equality of incorrigibility, we are left to wonder whether a victory that ensures the intellectual respectability and vitality of Christian theology at the price of lumping astrology, faith healing, nuclear physics, and history in the same epistemological bin might not be a Pyrrhic one.

Such an interpretation of Holmer's writing strongly supports the view that confessional method in theology is "fideist" and "internalist," if not an outright rejection of any standard of rationality between disciplines or language communities. But this interpretation, although understandable due to a preponderance of evidence, ignores other observations that Holmer makes about the problem of the relation of theology to other intellectual disciplines.

For example, he cites transubstantiation and "real presence" as theological concepts from which "the life seems to have gone," and which take "historical imagination" even to see the point of.[12] Whereas Holmer considers it misguided to focus on such concepts and cases and to generalize that all are similarly historical in character, he does allow that some relatively marginal theological concepts are subject to the currents of historical change. But most are not like that, he seems to think; most are "immutable" and "timeless."

Holmer cites an example to demonstrate that concepts are not theory-laden. He posits a visiting Martian who, aware that Earthlings are Copernican in their astronomy, is initially confused at the prevalence of the words "sunset" and "sunrise" in their ordinary speech. Holmer suggests that the Martian must learn to get language games straight and recognize that these terms were "not really theory-words nor theory concepts every time they were used. Insofar as they called attention to very striking happenings, they still do so, Copernicus notwithstanding" (TC, 149–50). In other words, such concepts have "a striking nonhistorical character about them" (TC, 150).

It can be argued, however, that "sunrise" and "sunset" once fit perfectly into a theoretical account of the "striking happenings" attendant to the apparent proximity of the sun to the horizon, that is, that the sun "came up" over the edge of the disk of the flat earth and "went down" below it at the end of the day. But the terms have been purged of all theoretical meaning, because the theory in which they were set has become intellectually outmoded. In other words, they were once theory words, but are no longer. The same transition occurs in the use of terms such as "real presence" or "transubstantiation." There once were elaborate theories, drawing on Aristotle's metaphysics or claims about the spatial location of the resurrected Christ, into which these terms fit as genuine theoretical terms. The words are still in use, even though few find those overarching theoretical schemes plausible (for reasons having to do with the *historical* decline of the theories on which they rested). And when the words are used, they call attention to "striking" features of the attitude of Christian congregations toward a central ritual of their faith. Holmer's account does not explain why such a term as transubstantiation should drop out of use, particularly if "theology proposes something eternal" (WTID, 12) and if its rationality is determined solely by "distinctively religious categories" (SLLR, 65).

Holmer's preoccupation is with settling problems of the meaning of language, not to determine whether what the language says is true. He rightly draws attention to the need to listen closely to the terms used by speakers in a particular religious community or linguistic "subcommunity" to determine the meaning of the terms. He does not, however, take a stand on whether the meaning of terms should have reference to the state of the real world. Nevertheless, he recognizes the distinction. He says, for example:

> A concept is learned by mastering the way the word is used ... and therefore the concept is more like a rule than a thing. . . . Thus, theologians are in no position to justify the processes by which words have acquired their meanings. For justification is entirely out of order here. . . . These circumstances of social and religious life were not invented, and neither were the meanings that grew with them. If religious words made great differences, they were also very meaningful. But all that was said with those meaningful words is another matter altogether. Religious men thought they were speaking truly about a world that was created and a God who died for the sake of the world. These beliefs indeed suppose religious concepts; for they make statements that tell us what is the case, and therefore they must refer. But it is one thing to see that language, even religious language, has meaning; it is another thing to see that it is true.[13]

Holmer does not follow up on this observation. He correctly insists on the social and historical location of the meaning of concepts and language and on one's need to attend to the ways in which "inside" speakers of a distinctively religious language use it if one is to discover meaning. Unfortunately, he overgeneralizes in claiming that "rationality is polymorphic" (SLLR, 73–74) and oversimplifies in describing terms from different language-games as hereby "incommensurable."

This lack of attention to reference and truth distinguishes some other forms of confessional theology most clearly from Holmer's. Confessional theologians generally will agree that our routes to knowledge of the world are polymorphic because of the irreducible cultural and historical particularities that we bring to even our most fundamental perception and valuation. More nuanced versions of confessionalism will also insist that all knowledge is finally concerned with the "one beyond the many," the reality that lies behind and beyond the inevitable relativity of perception.

Holmer's position strongly justifies the charge that confessional theology is fideist, if not obscurantist, and that the major goal of confessional method is to immunize theology from not only correction by, but even interaction with, other intellectual disciplines or cultural changes. But this is not a necessary feature of confessional method, even among those who draw their epistemological inspiration from Wittgenstein. The thought of D. Z. Phillips substantiates that assertion.

D. Z. PHILLIPS'S CONFESSIONALISM

Phillips, primarily a philosopher rather than a constructive theologian, is concerned to analyze religious speech as it is actually found rather than to make proposals about preferable use.[14]

As in Holmer, the central concern of Phillips's work is the rejection of "foundationalism." And, again as in Holmer, the key to this rejection is an insistence on diverse criteria of rationality which vary depending on the context of the discourse in question. Phillips states the issue thus:

> It has been far too readily assumed that the dispute between the believer and the unbeliever is over a matter of fact. Philosophical reflection on the reality of God then becomes the philosophical reflection appropriate to an assertion of a matter of fact. . . . A necessary prolegomenon to the philosophy of religion, then, is to show the diversity of criteria of rationality; to show that the distinction between the real and the unreal does not come to the same thing in every context.[15]

Much more acutely than Holmer, Phillips argues that this is the case not because of some desire to protect religious utterance from the incursions of other kinds of discourse upon it, but because of the nature of religious language and belief itself.

Religious language cannot be placed on the same "scale" as other language—extrareligious reasons for such belief cannot be provided—by virtue of its absolute character, its characteristic of pointing beyond the ordinary criteria of meaning and truth of empirical discourse.[16]

41

Such absolute beliefs are not "testable hypotheses," as are empirical propositions, but rather are absolutes for believers in that they "predominate in and determine much of their thinking" (RBLG, 90). In this regard, fundamental religious propositions are not accepted on the basis of proof of any kind; rather, they provide a "world-picture," an overarching and indemonstrable framework within which the believer thinks and responds to experience.[17]

Within any language-game, there are fundamental propositions that, although not demonstrated, provide the foundation for sensible discourse within that realm. Phillips develops an example drawn from the realm of discourse about the empirical world: the assertion "This is a tree" uttered by a speaker standing in front of a tree. While the denial of such an assertion is imaginable in certain contexts (in a philosophy class, for example), serious denial of such propositions in everyday life situations would indicate that the speaker was insane or had not mastered the usage of ordinary empirical object discourse.

Fundamental religious propositions, Phillips argues, share certain logical and grammatical features with such empirical propositions. Among the features of all such propositions, Phillips identifies six essential ones: (1) such propositions are free from the possibility of error, since they themselves provide the standards for the assessment of error; (2) they are not hypotheses, but rather presuppositions; (3) they are not based on experience, but rather "show themselves" in experience by being "brought to" experience and providing the means of organizing the experience one has; (4) they are rarely explicitly formulated, but rather are reflected in the ways in which we formulate propositions in the ordinary course of activity in that "form of life"; (5) when people in our culture are cut off from the propositions we take for granted and do not question, we describe them as joking or insane, not as "mistaken" since "mistakes" only make sense within a shared framework; and (6) these propositions have the status they do, not because of anything inherent in their nature, but rather from the presumptions of the culture within which they function. Correlatively, changes in the culture will bring about changes in the norms of reasonability that function within that culture and, hence, in the propositions that enjoy this status.[18]

These beliefs, then, are "taught not as beliefs which require further reasons to justify them. They are not opinions or hypotheses. It is this feature of such beliefs which make it odd to say that those who believe them hold opinions" (DGE, 164). From this it follows that the logic of their denial is also peculiar. The denial of empirical propositions, such as the claim that it will rain tomorrow, is a straightforward dispute about empirical matters in which both speakers share the same set of terms and criteria for verification and disagree only about the factual character of a future event. But the person who denies the religious proposition such as the Last Judgment is not making an analogous denial. Rather,

> It is not a question of his believing the opposite of what the believer thinks, but of his not sharing or having the belief at all. If one characterizes the lack of belief as believing the opposite one falsifies the character of belief and the character of non-belief. On this view, the affirmation of faith would become belief in an hypothesis or matter of fact which may or may not be the case. We have already seen, however, that religious beliefs have an absolute, necessary, character. But this view also falsifies non-belief, for it makes it look as if the nonbeliever shares the same mode of discourse but makes the opposite judgment within it. Whereas the non-believer does not participate in that religious mode of discourse. (DGE, 166)

In the case of an empirical proposition (for example, the utterance "This is a tree" made while standing in front of a tree) the serious denial of such a proposition cuts the speaker off from rationality itself. That is, the proposition in question is essential to rational discourse: it is a paradigm case of intelligible utterance. This is so because empirical propositions enter into other propositions that run through multifarious activities and situations, and generally constitute the essential material for discourse (DGE, 167). In other words, the "empirical identification" language-game is a pervasive "subgame." It crosses so many realms of discourse that inability to use or comprehend it would virtually render one incapable of rational discourse in any "game."

Denial of a religious proposition does not, however, similarly compromise one's grasp on rationality itself. If one denies belief in the Last Judgment, for example,

> One could be said to be cut off from a way of thinking or from a certain perspective on life. Yet, one cannot, as in the case of the other propositions, say that the person who is cut off from religious beliefs and perspectives is cut off from reason. . . . Therefore one can imagine a person having no use for [religious concepts] and still being able to share a common life with other people, whereas a person having no use for [the fundamental empirical propositions] will have cut himself off from such a life. (DGE, 167)

Another important difference between empirical and religious fundamental propositions is a function of the existence of genuine alternatives to the proposition in question. There are no genuine alternatives within the realm of rationality to the assertion "This is a tree" when one is standing in front of one. Its denial simply indicates that one has not learned the minimal rules for the usage of words or the grammar of the language. In the case of religious assertions, however, there are alternatives.

> One cannot say that there is no alternative to "God created the heaven and earth," since "I don't believe it" or "I don't accept that" expresses such alternatives. These are genuine alternatives since they indicate that the person has no use for the religious belief, that it means nothing to him, that he does not live by such a belief, or that he holds other beliefs which exclude religious faith. In this latter case, however, the alternatives are not alternatives within the same mode of discourse, but rather, different perspectives on life. (DGE, 168)

Theology is confessional, for Phillips, in the sense that only those who share a common perspective on life are capable of theological inquiry, which is by definition the stating of "what can sensibly be said of God" *within* such a shared tradition.[19]

How is it, then, that individuals come to share such a perspective, such a common religious discourse?

> This is taught to children through stories by which they become acquainted with the attributes of God. . . . [T]he child does not listen to the stories, observe religious practices, reflect on all this, and then form an idea of God out of the experience. The idea of God is being formed in the actual story-telling and religious services. To ask which came first, the story-telling or the idea of God, is to ask a senseless question.

> Once one has an idea of God, what one has is a primitive theology. . . .
> [T]heology decides what it makes sense to say to God and about God.
> In short, theology is the grammar of religious discourse. (PTRG, 5–6)

The crucial question in confessional theology generally and, as we have seen in the earlier discussion of Holmer's view, in Wittgensteinian philosophy of religion in particular, is: To what extent and in what way is religious discourse in interaction with, or corrigible by, alterations in other realms of knowledge and discourse in the common culture? Or, to note a slightly different but related problem, in what ways can religious discourse be erroneous? Does the emphasis on the lack of foundation to religious belief, and the insistence on the separate language-game within which alone theological utterance gains meaning, function for Phillips (as it tended to do for Holmer) to immunize religious discourse from "external" criticism or, indeed, from interaction with other realms of discourse?

Phillips is much clearer than Holmer on this set of questions, indicating two distinct ways in which religious discourse can be in error: it may commit "blunders" within the language-game of a particular religious view of life (by becoming superstitious or metaphysical, for example), or it may fail in its interaction with other forms of life and realms of experience. Each of these modes of failure warrants its own treatment.

First, there are mistakes that only make sense within a particular language-game. Different religious traditions set the limits and define the things that can sensibly be said about God. But it is a mistake, Phillips argues, to think that there is some general set of criteria for assertions that may be made about God. Rather, "the criteria of meaningfulness cannot be found *outside* religion, since they are given by religious discourse itself. Theology can claim justifiably to show what is meaningful in religion only when it has an internal relation to religious discourse" (PTRG, 4).

At times, mistaken uses of the language occur; these blunders are corrected by "native speakers" who are familiar with the proper usage of that language. There are no criteria for such blunders external to that language, however, such that an a priori linguistics can dictate the bounds of proper speech for all languages, independently of empirical analysis of the particular language in question.[20] Similarly, there are no a priori criteria for legitimate religious discourse, but only the bounds of sense internal to a given religious community (RBLG, 92–93). For example, for an orthodox Christian, the statement "God has a material body" is impermissible, not because it is empirically falsifiable, or necessarily inappropriate in the religious discourse of all communities (as evidenced by the fact that the Mormon conception of deity is by definition material), but simply that the Christian concept of God includes the idea of immateriality.

Phillips, unlike Holmer, displays awareness that the fact that religious utterances are given meaning within language-games does not mean that religious language is necessarily secure from apparent falsification or decline of meaning for the believer. The interesting question, he insists, is not whether beliefs are immune from such decline, but how we are to describe their erosion. In order to do that, one must first describe the nature of the beliefs in question. Phillips writes:

> If . . . having heard people praising the Creator of heaven and earth, glorifying the Father of us all, feeling answerable to the One who sees all, someone were to say, "But these are only religious perspectives, show me what they refer to," this would be a misunderstanding of the grammar of such a perspective. . . . The religious pictures give one a language in which it is possible to think about human life in a certain way. The pictures . . . provide the logical space within which such thoughts can have a place. When these thoughts are found in worship, the praising and the glorifying does not refer to some object called God. Rather, the expression of such praise and glory is what we call the worship of God.[21]

If religious beliefs function that way, how are we to describe their loss? Should we say that the religious picture has been decisively engaged and refuted on its own ground by information coming from "outside" the religious perspective? Not in the first instance, at least, Phillips argues. Rather, we would say that "the attention of the individual has been won over either by a rival secular picture, or, of course, by worldliness, etc." (RBLG, 115).

The discussion so far shows Phillips to be in substantial agreement with Holmer that religious belief is hermetically sealed off against other perspectives drawn from other aspects of the culture, that, as Holmer put it, there is a "divine magisterium" retained *in toto* or not at all. Precisely in this respect, despite the common orientation and source, the ways of Phillips and Holmer part.

For one thing, Phillips notes (as Holmer was constrained by factual necessity to mention, and then hastened to attempt to deny) that theology does undergo change in response to changing historical circumstances. In fact, although he cites important differences as well, he posits ways in which the evolution of theology within a tradition can be seen as analogous to the changes in scientific theory over time (PTRG, 6–9).

How can we account for that bit of historical fact? Why should theology undergo such change? If the meaning of religious discourse is dependent on a community that finds that meaning, why should it be subject to change, particularly if the meanings are communicated through the telling of stories to children at an early age? We must consider that religious meaning and activity are not cut off from the full range of other activities in life, but rather interact with and interpenetrate them:

> If we think of religious worship as cut off from everything outside the formalities of worship, it ceases to be worship and becomes an esoteric game. What is the difference between a rehearsal for an act of worship and the actual act of worship? The answer cannot be in terms of responses to signs, since the responses to signs may be correct in the rehearsal. The difference has to do with the point the activity has in the life of the worshippers, the bearing it has on other features of their lives. . . . So far from it being true that religious beliefs can be thought of as isolated language-games, cut off from all other forms of life, the fact is that religious beliefs cannot be understood unless their relation to other forms of life is taken into account. (RBLG, 97)

The important task for us is to characterize accurately the relation between religious beliefs and these other forms of life. Phillips is aware of the fears of Kai Nielsen and others that "Wittgensteinian fideism" simply immunizes religious discourse from criticism, that it " 'saves' religion only at the cost of leaving the door open to any sort of transcendental metaphysics—and indeed to superstition and nonsense of the most arrant sort."[22] This recognition leads him to acknowledge certain weaknesses in the "game" analogy:

> The point of religious beliefs, why people *should* cherish them in the way they do, cannot be shown simply by *distinguishing between* religious beliefs and other features of human existence. . . . The importance of religion in people's lives cannot be understood simply by distinguishing between religion and other modes of social life, although, as we have seen, there are important distinctions to be made in this way. . . . [I]f religion were thought of as cut off from other modes of social life it could not have the importance it has. (RBLG, 93–94)

. . . .

> Wittgenstein takes it for granted that the same language [in the ordinary sense of "language," i.e., English, French, etc.] is being spoken in the different language-games. But if this is so the sameness or unity of that language cannot be explained by describing the way in which any *particular* language-game is played. The problem becomes acute when Wittgenstein says that each language-game could be a complete language in itself. One reason why Wittgenstein said that each language-game is complete is that he wanted to rid us of the supposition that all propositions have a general form. The different language-games do not make up one big game. . . . [B]ut [this assertion] gives rise to new problems. The different games do not make up a game, and yet Wittgenstein wants to say that a language, the same language, *is* a family of language-games—that is, that this is the kind of unity a language has. At this point there is a strain in the analogy between language and a game. (RBLG, 94–95)

. . . .

> The specific limitation of the game language is that: Religion has something to say about aspects of human existence which are quite intelligible without reference to religion: birth, death, joy, misery, despair, hope, fortune and misfortune. The connection between these and religion is not contingent. A host of religious beliefs could not be what they are without them. The force of religious beliefs depends, in part, on what is outside religion. (RBLG, 97)

What is the nature of this dependence? Religious responses are connected to other realms of discourse; they must not be "fantastic," and they become so if they "ignore or distort what we already know" (RBLG, 98). But the issue remains clouded when we inquire more deeply into the nature of these purported "facts" that we already know and with reference to which religious belief may distort our perception. What kinds of facts are these?

To illustrate the ways in which religious discourse could contradict what we already know, Phillips considers typical religious activities such as a mother's prayer to the Virgin Mary for her child, or a boxer's crossing himself before a fight. He asks whether such activities necessarily imply belief in the causal efficacy of such practices. Phillips argues that they are not genuine religious activities at all, but "blunders" or "superstition." He writes:

> What characterizes the superstitious act in this context? Firstly, there is the trust in non-existent, quasi-causal connections: the belief that someone long dead called the Virgin Mary can, if she so desires, determine the course of an individual's life, keep him from harm, make his ventures succeed, and so on. Secondly, the Virgin Mary is seen as a means to ends which are intelligible without reference to her: freedom from harm, successful ventures, etc. In other words, the act of homage to the Virgin Mary has no importance in itself; she is reduced to the status of a lucky charm. (RBLG, 103–04)

Two distinct, although related, points emerge here. First, such activities contravene our knowledge of the nature of causality and, to that extent, contradict what we already know, as does the belief that the stars determine one's destiny or that a rabbit's foot alters one's prospects for good fortune. Second, such activities are blunders precisely because they move religious discourse into the same sphere as empirical causality. In other words, because such activities compromise or negate the autonomy of religious discourse with reference to other modes of discourse, they must be blunders within the religious language-game.

The most difficult issue in Phillips's thought resides in the second point. He claims to be a philosopher, not a theologian. That is, the goal of his analysis is ostensibly to clarify religious activity as it actually exists rather than as it ought to be. Quoting Wittgenstein, he states, "Philosophy leaves everything as it is."[23] But unless he can show that religious believers within a particular tradition agree that a given understanding of an activity is, in fact, a blunder, he can be correctly charged with importing a normative claim into his supposedly purely descriptive analysis. It is self-evident that many religious believers think that such activities do indeed have causal efficacy. How can Phillips handle this?

His main recourse is to show that, although religious believers make utterances that seem linked to empirical expectation, those utterances do not obey the same rules of logic and expectation that such propositions ordinarily would if they really were empirical. For example, when a believer says that God's love is eternal, he or she is not predicting that as a matter of fact there will not be any future time at which God will not be loving. Such a prediction would be subject to empirical falsification in exactly the same way that the claim that one will love one's wife for the whole of one's life is: it "depends on how things go, it may change, it may end in failure."[24] The point is not that the love of God is truly eternal, but that it remains constant as long as the believer remains a believer. As Phillips puts it, "This is the love of God, the independence of which from what happens is closely bound up with the point of calling it eternal."[25]

Of course, as a matter of empirical fact, experiences such as sudden illness and death of loved ones lead some persons who previously would have described themselves as believers to say, "If God allows 'x' to happen to me, then I can no longer believe in God." Often such persons are thereupon described as having had an inadequate or immature belief in God or as having understood religion superstitiously.[26]

In short, religious beliefs are not free from contact with more general realms of experience. Indeed, they interpret that experience in the largest possible context for the believer. Yet they observe the rules of a unique logic that distinguishes them from other kinds of language and that requires the use of a different (and unique) set of tools in their analysis.

Phillips, then, is significantly more nuanced in his treatment of internalism and related matters than was Holmer. He recognizes both the historical location and change of religious traditions, and also that religious utterance interacts significantly with the full range of human experience in a way that Holmer's discussion tends to obscure. He shares with Holmer an emphasis on the logically peculiar status of religious claims deriving from Wittgenstein's basic program. But he develops it in a manner that makes him substantially less culpable on the charge of importing into ostensibly philosophical analysis a strong normative theological agenda.[27]

LINDBECK'S DESCRIPTION
OF RELIGION

An important new perspective on theological method has recently been proposed by George Lindbeck in his *The Nature of Doctrine*.[28] He claims that the "experiential-expressivist" and "dogmatic-proposition-al" methods have dominated in modern theology, and then proceeds to advocate an alternative, "cultural linguistic" methodology as more defensible and fruitful for contemporary constructive theology.

Lindbeck's proposal bears a close resemblance to the concerns and constructive positions of the family of confessional theologians considered herein, although his position is cast in quite different language. Similarly, many of the criticisms that have been leveled at Lindbeck's position parallel those made of confessional theology in general.

Lindbeck's project comprises two distinguishable enterprises: a descriptive theory of religion and religious thought, and an application of that theory to the normative question of the nature of Christian religious doctrine.[29] He attempts to assess the adequacy of the latter argument by using it to elucidate particular issues of doctrinal dispute in contemporary theology. Of greater significance for the scope of this book, however, is the first facet of Lindbeck's work: the general theory of religious doctrine he develops and his view of the nature of theological constructions which is linked to that theory.

Lindbeck begins with a critical analysis of the state of contemporary understanding of the nature of religious doctrine. He argues that much of that understanding is founded in dichotomy between an "experiential-expressive" view, which sees doctrines as "noninformative and nondiscursive symbols of inner feelings, attitudes, or existential orientations" (*ND*, 16), and a dogmatic-propositional understanding, according to which doctrines are "informative propositions or truth claims about objective realities."[30]

Lindbeck's fundamental claim is that such a division of disputed territory hinders comprehension of what actually occurs in doctrinal development and argumentation, especially in ecumenical contexts. How are we to account, he asks, for the apparently bizarre result of

51

ecumenical discussions in which traditions that centuries before did literal and theological battle over doctrinal differences, announce agreement between their positions without granting that either has capitulated to the other? According to the cognitivist understanding, such claims are necessarily false: Agreement is only possible if one or both sides abandons its earlier view. For the experiential-expressivist view, on the other hand, such claims are possible if the traditions in question discover that, beneath their differences in language and formulation, both doctrinal positions are attempts to articulate fundamentally equivalent "experiences" (*ND*, 16–17).

Arguing that neither of these analyses accurately captures the nature of such ecumenical exchanges, Lindbeck proposes an alternative view of doctrine as necessary for descriptive accuracy. According to this cultural-linguistic view, church doctrines are "rules" governing the speech and conduct of Christian communities (*ND*, 18).

This change is significant because the rules retain invariant meaning while varying in their applicability to a variety of contexts. As an example, Lindbeck cites the rules "Drive on the right" and "Drive on the left." The meaning of the rules is both clear and contradictory, but only one is binding, depending on the context of driving in Britain or in the United States (*ND*, 18).

Lindbeck also argues that the dominance of the cultural-linguistic approach to religion in nontheological religious studies warrants a careful assessment of its applicability to theological contexts as well. Although he notes the cultural and intellectual factors that apparently make adoption of the cultural-linguistic approach undesirable or difficult (*ND*, 19–25), he also observes the "growing gap between theological and nontheological approaches," and that the gap "tends to ghettoize theology and [deprive] it of that vitality that comes from close association with the best in nontheological thinking" (*ND*, 19).

In developing a theory of religion and corresponding theory of doctrine, Lindbeck seeks to elaborate a view of theology which draws freely on the resources of the dominant "religious-studies" approach and overcomes the methodological gap he considers characteristic of the rival methods. As he writes: "The theory of religion and religious doctrine that [this book] proposes is not specifically ecumenical, nor Christian, nor theological. It rather derives from philosophical and social-scientific approaches; and yet . . . it has advantages, not only for the nontheological study of religion but also for Christian—and perhaps also non-Christian—ecumenical and theological purposes" (*ND*, 7–8).

The degree to which Lindbeck justifies his claims for the superiority of the cultural-linguistic approach is debatable.[31] That does not detract, however, from the value of examining his thought. Lindbeck formulates the fundamental question that distinguishes his position from experiential-expressivist views as follows: "Whether it is conceptually and empirically better to picture religions in expressivist fashion as products of those deep experiences of the divine (or the self, or the world) which most of us are accustomed to thinking of as peculiarly religious, or whether one should opt for the converse thesis that religions are producers of experience" (*ND*, 30).

In contrast to experiential-expressivist models of doctrine according to which experience is logically and temporally prior to doctrinal expression, the cultural-linguistic approach reverses this order.[32] It stresses the ways in which experience is shaped, formed, and led by the categorical scheme transmitted by religious enculturation:

> A religion can be viewed as a kind of cultural and/or linguistic framework or medium that shapes the entirety of life and thought. . . . Like a culture or language, it is a communal phenomenon that shapes the subjectivities of individuals rather than being primarily a manifestation of those subjectivities. It comprises a vocabulary of discursive and nondiscursive symbols together with a distinctive logic or grammar in terms of which this vocabulary can be meaningfully deployed. Lastly, just as a language . . . is correlated with a form of life, and just as a culture has both cognitive and behavioral dimensions, so it is also in the case of a religious tradition. Its doctrines, cosmic stories or myths, and ethical directives are integrally related to the rituals it practices, the sentiments or experiences it evokes, the actions is recommends, and the institutional forms it develops. (*ND*, 33)

This sort of claim is a familiar one. It is a common aspect of all forms of Wittgensteinian analyses of religious experience and community. And the "theory-laden" character of perception has been, as evidenced earlier, a hallmark of the newly developed historicist philosophy of science of this century.

Like the confessional theologians examined so far, Lindbeck emphasizes that the meaning of religious terms and ideas is to be found not in some other, more neutral, language from which they derive their meaning but rather from within the realm of religious discourse and community. Lindbeck writes, "Religions are seen as comprehensive interpretive schemes, usually embodied in myths or narratives and heavily ritualized, which structure human experience and understanding of self and world" (*ND*, 32).

As was stressed by Holmer and Phillips, so too for Lindbeck, becoming religious is something quite different than affirming propositions to be true: "To become religious . . . is to interiorize a set of skills by practice and training. One learns how to feel, act, and think in conformity with a religious tradition that is, in its inner structure, far richer and more subtle than can be explicitly articulated" (*ND*, 35). The noncognitive, aesthetic, and symbolic components of religion are, on this analysis, not secondary or derivative. Rather, they communicate the nonrational but essential skills that form the essence of the religion. Once formed, these skills enable the practitioner of the religion to discriminate intuitively between valid and inappropriate, tasteless, or mistaken objectifications of the religion.

This is the major contrast between the experiential-expressivist and the cultural-linguistic view of the relation between experience and expression, Lindbeck argues. The experiential-expressivist imagines prethematized experience to be the fundamental feature of religious experience, and the symbolic and linguistic formulation to be the dependent variable, always partially inadequate. In contrast, the cultural-linguistic approach sees the "expressive and communicative symbol systems" to be primary. They provide the schema through which experience is mediated to the believer, even in unreflective experience (*ND*, 36).

Although granting that profound personal religious experience is necessary to originate and energize a religious tradition, Lindbeck stresses the priority of the communally shared language and symbols that assign meaning and significance to those experiences. He writes, "First come the objectivities of the religion, its language, doctrines, liturgies, and modes of actions, and it is through these that passions are shaped into various kinds of what is called religious experience" (*ND*, 39).

On this analysis, therefore, religions are not essentially diverse symbolizations of a common core religious experience. Rather, "religion" is "a class name for a variegated set of cultural-linguistic systems that . . . differentially shape and produce our most profound sentiments, attitudes, and awarenesses" (*ND*, 40). Religious thought, as the "explicitly formulated statements of the beliefs or behavioral norms of a religion" (*ND*, 35), is secondary. Ritual, prayer and the other defining practices of the religion, Lindbeck argues, are "normally more important" than those statements, and are at most "helpful in the learning process" undergone by a practitioner of the religion (*ND*, 35).

LINDBECK'S NORMATIVE CONCERN

As seen earlier, for both historicist philosophy of science and Wittgensteinian and Barthian analyses of religion, the question of the nature of truth is difficult and problematic, especially if one begins with a presupposed correspondence model of what truth must be. Judging from the degree of incommensurability of perspective presumed in such a case, views such as those examined here must either embrace relativism wholeheartedly (as in the case of Holmer) or else struggle to achieve a balance between perspectival variability and some account knowledge in general which permits judgments of better and worse among those perspectives. Lindbeck's view of theology is no exception to this problem. Indeed, the most telling criticisms of his view point to the scanty and inadequate treatment he affords to this crucial question.

The general nature of the issue is now familiar to us: If communities shape and form the perceptions of their members by means of pre- and nonrational rituals, symbols, and linguistic divisions, to what degree is intercommunal communication (not to mention adjudication of disputes) possible? If there are no community-neutral "observer languages," is the only account of knowledge possible irreducibly pluralistic? Since Lindbeck, like Holmer and Phillips, is concerned to emphasize the priority of the linguistic and communal particularity over the common and foundational, he too must declare himself on this range of questions.

In dealing with the problem of the truth of theological assertions, Lindbeck distinguishes three kinds of truth, each peculiar to one of the three theological methods he has elaborated earlier. First, traditional orthodox theological method understands theological assertions to be propositions, and true theological propositions to be those which correspond to ontological reality. Second, the experiential-expressivist model understands truth differently, with truth being "a function of symbolic efficacy" (*ND*, 47). That is, religious assertions have as their essential meaning the expression of nonlinguistic experience, and can be compared in their adequacy solely in the degree to which they do this well or poorly.

Third, the cultural-linguistic model advocated by Lindbeck sees religions as wholes, each constituting a "different [idiom] for construing reality, expressing experience, and ordering life" (*ND*, 47–48). Assessment of the "truth" of a given religious system involves two separate judgments. On the one hand, it is a matter of determining the religion's "categorial adequacy"—the degree to which the categories that form its idiom make it possible to "speak meaningfully of that which is . . . most important" (*ND*, 48). This level of assessment guarantees only meaningfulness, and not truth about the world, yet Lindbeck asserts that the adequacy of the categories alone warrants his referring to such religions as "categorically true religion" (*ND*, 48). On the other hand, since each religion constitutes its own idiom, there may be significant and unresolvable incommensurabilities among them "in such a way that no equivalents can be found in one language or religion for the crucial terms of the other" (*ND*, 48). This, unfortunately, makes it de facto uncertain whether it will be possible to determine which religions possess categorial adequacy, and which possess it in the highest degree.

The implication of this understanding of truth is potentially highly relativist. As Lindbeck writes, "Unlike other perspectives, [the cultural-linguistic approach] proposes no common framework . . . within which to compare religions" (*ND*, 49). If this were true, however, the cultural-linguistic view could not account for the fact that religions, perhaps especially Christianity, make the claim that their religious tradition and perspective are unsurpassably true?

Such claims are easily handled on the propositionalist account. The propositions of a given religion are unsurpassable if and only if they correspond absolutely, or as closely as possible given the limitations of human language, to the nature of things in themselves. For experiential-expressivism, unsurpassability is attributed to the absolute adequacy of the symbolic categories of the religion for expressing the core religious experience or feeling.

Lindbeck's account of the cultural-linguistic alternative, however, removes the discussion of unsurpassability from the level of the propositions, formulas, and symbols of religion, and relocates it on that of the actual living out of religious life guided by them. The "truth" of such religious systems, then, is assessed on the basis of the whole, which must be found to "correspond or not correspond to what a theist calls God's being and will." Indeed, "as actually lived, a religion may be pictured as a single gigantic proposition. It is a true proposition to the extent that its objectivities are interiorized and exercised by groups and individuals in such a way as to conform them in some measure in the various dimensions of their existence to the ultimate reality and goodness that lies at the heart of things. It is a false proposition to the extent that this does not happen" (*ND*, 51).

Drawing an analogy between the life prescriptions of religion and a map, Lindbeck argues that a map, no matter how accurately it may represent the terrain in question, becomes a "false proposition" to the extent that it is misread or misinterpreted, and a true one, despite distortions of scale and proportion, to the extent that it allows the traveller to move efficiently toward and find his intended destination (*ND*, 51–52). The point of the analogy is that the "categorically and unsurpassably true religion is capable of being rightly utilized, of guiding thought, passions, and action in a way that corresponds to ultimate reality, and of thus being ontologically (and 'propositionally') true, but is not always and perhaps not even usually so employed" (*ND*, 52).

To further complicate this analysis, Lindbeck introduces another set of distinctions between "intrasystematic" and "ontological" truth. Intrasystematic truth is the coherence of a statement with the whole body of stories, practices, and propositions that define a given religious system (*ND*, 64). Although apparently similar to the traditional idea of a "coherence theory of truth," it is important to see that Lindbeck intends to include a much broader range of materials with which the statement must cohere than that traditional category would. One example he gives makes the point well. "The crusader's battle cry 'Christus est Dominus,' . . . is false when used to authorize the cleaving the skull of the infidel (even though the same words in other contexts may be a true utterance). When thus employed, it contradicts the Christian understanding of Lordship as embodying, for example, suffering servanthood" (*ND*, 64).

The kind of coherence in question, therefore, is coherence with a system constituted "not in purely intellectual terms by axioms, definitions, and corollaries, but by a set of stories used in specifiable ways to interpret and live in the world" (*ND*, 64). Lindbeck asserts that coherence theory on the soil of a cognitive-propositionalist view, by contrast, "is unable to do justice to the fact that a religious system is more like a natural language than a formally organized set of explicit statements, and that the right use of this language ... cannot be detached from a particular way of living" (*ND*, 64).

Of course, Lindbeck recognizes that intrasystematic truth alone cannot guarantee the ontological reality of a single intrasystematic proposition. Coherent religious systems are possible, at least theoretically, which fail wholly to correspond to the nature of reality. Lindbeck wishes to claim, however, that the ontological correspondence to be sought is a feature not of isolated or individual propositions, but rather of the "form of life," the "way of being in the world" specified by the religion (*ND*, 64). The question to be judged in making such assessments is whether that form of life as a whole corresponds to "the Most Important, the Ultimately Real" (*ND*, 64).

Drawing on the notion of "performatory utterances," those which bring about a state of affairs instead of describing a preexistent state, Lindbeck argues that such religious systems can be assessed not as free-floating collections of propositions, but in the context of a life being lived out under them. "One must be, so to speak, inside the relevant context; and in the case of religion, this means that one must have some skill in *how* to use its language and practice its way of life before the propositional meaning of its affirmations becomes determinate enough to be rejected" (*ND*, 68).

The upshot of these claims is that the question of the ontological correspondence of religious systems to reality is located not at the level of technical or academic theological assertions, but rather in the first-order language and practice of the religion. Such technical theology is, to return to the grammar analogy, noninformative about reality. Rather, it defines the grammar of properly religious speech. And even properly religious speech purports to make ontological claims, Lindbeck asserts, only when it is used in "speaking religiously," "when seeking to align oneself and others performatively with what one takes to be most important in the universe by worshipping, promising, obeying, exhorting, preaching" (*ND*, 68).

This is the essence of Lindbeck's attempt to offer an epistemology that maintains the crucial concerns of cultural-linguistic method in theology. Many issues crucial to larger concerns are, unfortunately, left vaguely or inadequately addressed, despite the complex distinctions drawn and the apparent efforts made to deal with them.

The essence of Lindbeck's position is summarized in the intent to do theology "intratextually" (*ND*, 113ff.). That is, he limits the method of theology to elaboration of a world view produced from *within* the symbol system of a given religion. "For those who are steeped in [the canonical writings of religious communities], no world is more real than the ones they create. A scriptural world is thus able to absorb the universe. It supplies the interpretive framework within which believers seek to live their lives and understand reality" (*ND*, 117).

Although this fundamental claim is common to the full range of confessional theological positions, it lacks subtlety on the connection between this intratextual view of the world and the many other views that all modern persons necessarily carry with them as well.[33] As one sympathetic critic has correctly observed: "In the full knowledge that we possess no Archimedean point outside cultural-linguistic systems, some of us nonetheless insist upon the theological imperative to discover and build upon what might be called 'intertextual' correspondences. And that not only for our own intellectual satisfaction, or to assure ourselves of the 'truth' of Christianity, but [for] the sake of the life of the church."[34]

In the absence of some accounting for those connections, despite the degree to which Lindbeck argues that his cultural-linguistic method harmonizes with nonreligious approaches to the study of religion, one necessarily suspects that the ultimate goal is the kind of confessional fideism often charged to all confessional methods. Indeed, in Lindbeck's case, one feels by the end of the work that the substantive theological goal is a defense of an essentially Barthian confessional position in the guise of Wittgensteinian philosophy of religion. David Tracy cleverly and accurately states the point: "Lindbeck's substantive theological position is a methodologically sophisticated version of Barthian confessionalism. The hands may be the hands of Wittgenstein and Geertz but the voice is the voice of Karl Barth."[35]

> Even those who agree that a "purely neutral" theory of rationality is never "purely neutral" and who will agree that skill, practice, etc., are crucial ingredients in any attempt to assess rationally all theological claims will remain unpersuaded that Professor Lindbeck's "epistemological realism" is other than relativism with a new name or that his "cultural-linguistic" grammatical model for theology is other than confessionalism with occasional "ad hoc" apologetic skirmishes.[36]

In the last analysis Lindbeck's more sophisticated view lacks what Holmer's lacked: an assessment of the relationship between the perspective and vision unique to the confessional tradition and those perspectives generated by other historical and social traditions, no less particular, and yet equally a part of the world view of every thoughtful modern person. Fundamentally, Lindbeck lacks an account of *"mutually critical* correlations of an interpretation of the meaning and truth of the tradition and the interpretation of the meaning and truth of the contemporary situation."[37]

FIORENZA'S FOUNDATIONAL THEOLOGY

Francis Schüssler Fiorenza has recently contributed significantly to the debate on theological method and to doctrinal theology.[38] Fiorenza's work is distinctive in the context of the positions considered here because first, it is based squarely within the Roman Catholic context of debate over "fundamental" or "foundational" theology, and second, it is informed by a wide range of Anglo-American and continental philosophical traditions.

Fiorenza also distinguishes himself by tackling substantive questions of doctrinal theology in the context of working out an articulation of theological method. His *Foundational Theology* deals both with questions of theological method and with a contemporary constructive theological statement regarding the resurrection of Jesus, the foundation of the church, and the church's mission.

Fiorenza situates his discussions of method firmly within the context of Roman Catholic discussions of fundamental theology. While this discussion can at times be arcane to the non–Roman Catholic reader, the issues at stake are the common fare of all Christian theology concerned with foundational issues: What kinds of truth claims can be made by theology, and what grounds or locates those claims?

Classically, Fiorenza notes, fundamental theology depended on a correspondence theory of truth. The claims of fundamental theology on this model are understood as factual claims about reality as it actually is, warranted and demonstrated by the extrinsic warrants of miracle and revelation and to some degree by intrinsic features of the revelation itself such as the sublime character of the message.

Fiorenza criticizes this model of theological method on a number of grounds. First, it "cannot deal with the normative, expressive, and existential dimensions" of religion—it is unable "to express what is not descriptive and propositional" (*FT*, 273). Second, it is also inadequate in that it neglects the inherently interpretive aspect both of the "reality" being described in theology and of theological discourse itself—hence, it has a deficient sense of the nature of interpretation and the interplay between interpretation and experience.[39]

In light of the difficulties of the traditional method, more recent theology has adopted the method of correlation. In this approach, phenomenological analysis reveals fundamental structures of human experience as religious. Then the content of Christian revelation is shown to correlate with and meet those structures with appropriate "answers." Although Fiorenza analyzes various forms of this fundamental approach and recognizes differences among them, he also notes that all seem to assume a coherence rather than a correspondence theory of truth.[40]

Although Fiorenza treats the significant differences in method among Peter Hodgson, Karl Rahner, and David Tracy, he judges them all as inadequate in their understanding of human experience. He writes, "they overlook the extent to which human experience and its theological interpretation is situated within the cultural tradition of Christianity and Western civilization" (*FT*, 281). So the experiential structures these methods purport to discover and with which Christian theology and revelation are supposed to correlate are, to a greater extent than the claims of the method allow, themselves conditioned by the theological and cultural tradition that they are supposed to validate.

Furthermore, Fiorenza also notes that the root structures of experience on which transcendental fundamental theology purports to build actually come closer to providing the ground of religious experience and considerably less of theology as a research discipline and intellectual task (*FT*, 281). He states, "Grounding Christian faith is not necessarily identical with grounding Christian theology" (*FT*, 281–82). In this respect, David Tracy's theology, which attends to the various "publics" with which theological language and community are in dialogue, is superior to some similar methods. It recognizes that the link between theology and common human experience is not direct but mediated, one in which experience is itself analyzed by some academic discipline that is then used and treated by theology (*FT*, 283).

The difficulty Fiorenza finds with all correlation methods is their lack of sufficient attention to the culturally and historically located character of experience and of thought. He claims that this "hermeneutical dimension" of both tradition and experience must be the distinctive contribution of foundational theology.

Fiorenza develops his own distinctive contribution to theological method by attending to the current debates regarding foundationalism in philosophical epistemology.

He draws the conclusion of those discussions for theological method:

> The critique of foundationalism questions whether fundamental theology can provide Christian faith and theology with an independent criterion and foundation.... Since the critique of foundationalism points to the cultural, social, and hermeneutical dimensions of "all foundations," it challenges attempts to limit hermeneutical retrieval of meaning to systematic theology and to the public of the Church in a way that would relegate the historical and transcendental analysis primarily to fundamental theology. (*FT*, 289)

It follows that any foundational theology that is possible in the present must avoid appearing to claim foundations in that stronger sense. Instead, it must be a more recursive intellectual enterprise than is implied in the metaphor of "foundation" built "up" from basic materials. Foundational theology must operate by means of a wide "reflective equilibrium" between reconstructing the meaning of Christian tradition and theology, "retroductive warrants" (those which are neither inductive nor deductive, but rather grounded in the illuminative inferences they make possible), and background theories (relevant theories from established nontheological disciplines which bear on the interpretation or meaning of specifically theological topics).

The foundational theologian must engage in critical hermeneutical reconstruction of Christian beliefs, practices, and convictions. Since religious traditions are diverse things, the theologian must engage in an interpretive act by which some elements are identified as more or less central. Placing this critical interpretation as the first task of foundational theology "presupposes that the starting point of foundational theology is neither historical facticity nor transcendental a priori" (FT, 305)—the failings Fiorenza had identified in classical fundamental theology and transcendental method, respectively. Instead, Fiorenza attends to the facts of historical particularity and diversity of religious tradition and does not claim more than is defensible about "common human experience." This beginning with historical diversity and particularity is, in fact, what warrants placing Fiorenza among the group of confessional theologians.

Retroductive warrants are Fiorenza's second essential element. By them, Fiorenza means to point to the degree to which religious beliefs prove in experience to be illuminating and to provide useful guidance to the believer. Fiorenza acknowledges the inherent circularity in such warrants: since they guide perception, much of what is experienced is experienced and expressed in terms set by the tradition. To this extent, religious beliefs function in a way analogous to the role of theory in guiding research and even perception in science. Fiorenza writes:

> Religious beliefs are subject to retroductive warrants in a manner analogous to other theories, insofar as a belief's ability to illumine experience and to guide praxis has in fact some independence from the belief's traditional coherence. . . . Since the hermeneutical retrieval of a tradition entails the retrieval of meaning and truth conditions, the ability of a hermeneutical retrieval to illumine and to guide praxis provides a warrant for the tradition. (FT, 308)

From the analysis of epistemology, Fiorenza notes that in principle there can be no escape from the inevitable hermeneutical circularity involved in the fact that one is biased to see things the way an inherited religious system disposes one to see them. Neither Liberation Theology's attempt to find a hermeneutically privileged class nor the Enlightenment's aspiration to a universal rationality can be looked to for assurance that one's present judgment of adequacy is correct in some absolute sense; no absolute sense if possible. Rather, reflective equilibrium balancing the present mode of seeing with the tensions that cannot be harmonized with it is the most adequate description possible of the way in which knowledge proceeds. "There does not exist a fulcrum independent of society's cultural tradition and experience that can provide a firm foundation" (*FT*, 309).

Finally, foundational theology must necessarily appeal to and be responsible to "background theories." They are theories from various disciplines that form the framework within which theological inquiry proceeds. In science, background theories provide the understanding of the functioning of experimental equipment and the meaning of the measurements they generate. Similarly, an assessment of the meaning and truth of Jesus' resurrection partly depends on "the nature of historical testimony, to the interrelation between literary form and content, and to narrative history and identity" (*FT*, 311).

For Fiorenza, therefore, the task and method of foundational theology involves the coordination of attention to these three distinct tasks and the claims they generate. Each is to some degree an independent issue or task, but must be balanced with the others in reflective equilibrium as the enterprise of theological construction proceeds.

The actual balancing of these competing concerns and issues in practice can only be seen properly in the doing of constructive theology. It is not essential that we assess Fiorenza's efforts in those areas, but we should note once again a position that is clearly confessional in its reliance on the historically and culturally particular character of religious communities, this time from a Roman Catholic thinker.

Fiorenza's emphasis on the essential tie of contemporary foundational theology to the historic traditions of Christianity also places him as a confessional thinker, concerned to claim a tie to the tradition as received (albeit interpreted by the contemporary theologian in light of a reconstruction of the essential claims of that historic tradition).

In addition, his emphasis on retroductive warrants for claims of theology—warrants which are looser than deductively or inductively rational—too places him in the company of confessional thinkers like the Wittgensteinians we have surveyed above.

But in Fiorenza too we find a position that clearly is not fideist. The importance of what Fiorenza labels background theories as an essential element in the reflective equilibrium he calls for clearly demonstrates an awareness of the importance of the connections between the claims of religious communities and theological thinkers on the one hand, and the "taken for granted" aspects of thought in all relevant intellectual disciplines outside the theological orbit, on the other. In this respect, too, we see a confessional method that maintains a balance and creative tension between the transmission of historically grounded claims and assertions and the openness and responsiveness to intellectual currents of its time and place.

CRITERIA FOR AN ADEQUATE CONFESSIONAL THEOLOGY

The survey of confessional methods just completed generates a range of tests for an adequate confessional theology. Those tests are the following:

1. Confessional method requires an adequate specification of the nature and grounding of the confessional aspects of its method. Ideally, this specification will be both intratheological—i.e., grounded in the nature and claims of theology itself—and extratheological and philosophical. Methods such as Barth's, which overtly at least offer only intratheological reasons for confessional method, have great difficulty dealing with inevitable and appropriate extratheological questions. On the other hand, methods such as Phillips's (which claims to depend only on an extratheological philosophical account of theological language) find it difficult and ultimately impossible not to import intratheological considerations into its analysis.

2. An adequate confessional method must explicity recognize that there is a complex interaction between the confessional perspective from which it begins and the perspectives provided by changes in the culture

and time in which it lives and in intellectual disciplines with which it is in dialogue. Nothing is gained and intellectual honesty is lost in attempting to deny this reality.

3. In light of the second criterion, an adequate confessional method is as clear as the subject matter permits about the ways in which theology determines the proper balance between confessional stability and response to perspectives and concerns that arise from those external perspectives. Here philosophical reflection on epistemology is helpful in explicating the ways in which theologians are to make decisions of inclusion, response, and rejection of extratheological claims and perspectives.

4. Lastly, and ultimately most importantly, an adequate confessional method will produce substantive theological work that receives the approbation of the confessing community as an intellectually adequate articulation of its confession.

3 | H. RICHARD NIEBUHR'S NONINTERNALIST PROPOSAL

H RICHARD NIEBUHR'S UNDER-
standing of method in theology is similar
in its confessionalism to the views of theological method already examined. Yet, it is less vulnerable to the charge of *internalism*—that
claims and assertions made within a language-game are not informed
by information or perspectives from outside that language-game—
than most of those positions. To elaborate the precise claims and
assumptions Niebuhr makes on epistemological issues, the grounds
for Niebuhr's claim that theology should operate in a confessional
manner need to be examined first.

THE RATIONALE FOR CONFESSIONAL METHOD

The dilemma of theology is powerfully stated by Niebuhr in the
following passage:

> I cannot think about God's relation to man in the abstract. This historical
> qualification of my relation to him is inescapable. I cannot presume to
> think as a Jew or a Mohammedan would think about God, though I
> recognize that they are thinking about the same God about whom I

think. Nor can I presume to rise above those specific relations to God in which I have been placed so as to think simply and theistically about God. There is no such being, or source of being, surely, as a Christian God (though there may be Christian idols); but there is a Christian relation to God and I cannot abstract from that.[1]

This illustrates the particular responsibility and situation of Christian theology as Niebuhr understands it. On the one hand, all theology must operate out of the particularities of the social and historical location of its author. On the other hand, all true theology, that which purports to speak of God, cannot be content simply to repeat the settled beliefs and assumptions of a particular subculture precisely because the God of whom it speaks is necessarily one and universal.[2]

This quotation further indicates the approach Niebuhr brings to the understanding of the role and nature of theology. Niebuhr offers a general philosophical account of human knowledge, reflection, and valuation that stresses the importance of the social and historical location of all thought, and into which Christian theology, confessionally understood, fits as species to genus. These epistemological claims are crucial for our purposes because they permit direct comparison with the claims of contemporary philosophy of science regarding the nature of intellectual development within the natural sciences.

Niebuhr's general philosophical arguments for confessional method are presented most fully in *The Meaning of Revelation*.[3] The themes of Niebuhr's general view of knowledge are pervasive in his work, however.

One major theme throughout Niebuhr's work is that of the social character of human thought and knowledge. From his earliest published book, *The Social Sources of Denominationalism*, to the posthumously published *The Responsible Self*, Niebuhr returns again and again to the social sciences as constitutive elements of theological construction. The discovery of the relativity of human cultures and points of view is, he asserts, the most characteristic feature of the modern view of the world (*MR*, 7ff.).

While appreciation and understanding of relativism is typical of modern thought in general, Niebuhr argues that the implications of relativism for the understanding of Christian theology and of the doctrine of revelation have not been fully explicated. Just as the interactional model for the description of the self has attained prominence in the social and biological sciences, so it must be applied to the analysis of the formation of religious communities and individuals. When so applied, Niebuhr asserts that even the most fundamental levels of "perception," whether religious or not, must be understood as arising from the influence of the social group surrounding the individual:

> Through the medium of language, with its names and categories, its grammar and syntax, its logic, I have been introduced to the system of nature, that is, to the system of nature as systematized by society. I classify the events and find their meaning in their relations to each other but do so always with the aid of the a priori categories of my social, historical reason, derived from my companions. To them I look not only for the categorial schemes with which I organize and interpret natural events but also for the verification of my reports of my direct encounters with nature. Hence it is that the concept of nature has a history and that men respond to natural events in varying ways in different periods of social history, on the basis of their different interpretations. (*RS*, 79–80)

Because of the recognition of this fundamental fact about human perception, Christian theology must be understood along strictly confessional lines: "Christian theology must begin today with revelation because it knows that men cannot think about God save as historic, communal beings and save as believers. It must ask what revelation means for Christians rather than what it ought to mean for all men, everywhere and at all times. And it can pursue its inquiry only by recalling the story of Christian life and by analyzing what Christians see from their limited point of view in history and faith" (*MR*, 31).

If theology must begin with "confession" of "what Christians see from their limited point of view in history and faith," it must do so not because it is able to demonstrate, on grounds that will be persuasive to external observers, the superiority of its point of view; indeed, Niebuhr argues that such attempts at "self-defense" are the most serious sources of error in theological thought (*MR*, x). Rather, theology must be confessional because there are simply no more neutral places from which to begin.

Because the most fundamental levels of perception are themselves conditioned and informed by images that are the inheritance of a particular history and social location, each self attempts to find meaning in its experience by assimilating new experience to the patterns of interpretation it bears from its past and its community. When that is not possible, the self modifies that pattern as necessary to account for the novel in its experience in ways that are continuous with those prior patterns. As Niebuhr expresses it:

> [The time-full self] comes to its meetings with the Thou's and It's with an a priori equipment that is the heritage of its personal and social past; and it responds to the action of these others in accordance with the interpretations so made possible.... Hence its responses in the present to the encountered Thou's and It's are guided largely by the remembered, a priori patterns. It seeks to interpret each new occasion by assimilating it to an old encounter, and it tends to respond to the newly present in the way it had learned to answer its apparent counterparts in the past. The responsive, interpreting self is highly conservative not because it loves the past but because its interpretative equipment binds it to the past. The categories of its historical reason largely determine what it can now know and how it will now respond. (*RS*, 96)

In short, it is because of this general fact about the historical character of human reason that Christian theology must adopt a confessional method. This is the case because "we are in history as the fish is in water and what we mean by the revelation of God can be indicated only as we point through the medium in which we live" (*MR*, 36).

Any theological method that claims to represent an alternative to confessionalism is misguided, therefore, because it fails to attend to the fundamental facts of human epistemology. Niebuhr quickly but incisively considers and rejects the proposed traditional alternatives to confessional method (natural theology, revelation theology that makes the highest possible claims for the Scriptures, and theology derived from individual experience)[4] arguing that all fail to acknowledge the ways in which our experience of nature, our reading of scriptural texts, and our personal religious experience are colored and shaped by the tools of interpretation we inherit from our tradition and our community (*MR*, 36–40).

The major reason for this failure on the part of other methods is that they fail to attend to the ways in which experience is shaped by valuation. Each seems to suggest that our experience, whether of nature, of Scripture, or simply of inner mystical or spiritual reality, is self-interpreting. Because of that assumed self-interpreting character, experience is believed to be capable of functioning as a nonconfessional "foundation" on which the edifice of Christian theology can be constructed. It is as a corrective to this error that Niebuhr elaborates his "relational theory of value" and his distinction between "inner" and "outer" history.

Niebuhr contrasts the natural sciences' aspiration to a "value-free" observation of the facts of experience with the tendency of "value theology" to remain "medieval," that is, "interested." By this he means that theology pursues knowledge and determines value with reference to immediate human valuations and preferences rather than to a sense of the place of things in a larger whole, a whole whose connection to human desires may be obscure.[5] So while he approves the general emphasis of theology since the nineteenth century on the question of value, he faults it for determining in advance what values are good or desirable (and, ostensibly, universal) and only subsequently identifying those values with Deity. In the sciences, by contrast, the acceptance of a relative theory of value (one that attends to the value of things with reference to the object of study rather than with reference to the observer) has made possible the kind of more objective knowledge of those objects that typifies the modern scientific approach (VTT, 97). And this approach to the knowledge of the natural world has not been incompatible with the persistent faith of the scientist in the objective reality of the objects of scientific study (VTT, 98–99).

A theology that begins its work with a system of values which are not themselves derived from the object of theological study (God), but from analysis of human nature or ordinary human valuations, will not be a theology that is independent and accurate in its description of the Divine. Niebuhr writes:

> Science to be worthy of the name requires independence. As a science it can be interested only in the accuracy of its observations and descriptions, not in their value from some point of view foreign to its own.... A theology that borrows its first principles from sociology, psychology or ethics has become a part of one of these sciences, and, since it has no independent point of view or object, the results of its researchers have been determined from the start. (VTT, 100–01)

Just as the progress of the natural sciences required the resolute rejection of the valuation of the objects of scientific study with reference to human interests for the development of a more objective method, so theology must avoid the assumption of systems of value which are constructed prior to the development of theological study and into which concepts of God are expected to be fit.

This analysis of the application of value theory to theology has direct implications for the understanding of the nature of revelation in theology and for the defense of confessional theology on general epistemological grounds. In all the forms of value theology that Niebuhr analyzes, from Ritschl to American realists such as Macintosh and Wieman, he asserts that there is a tendency to begin with "universally valid and absolute values, discovered apart from religious faith, and to . . . seek the essence of the religions in the common criterion of all religions" (VTT, 104). But this is inadequate, Niebuhr argues, because religions are "stubbornly individual" (VTT, 104). They are "directed toward the particular God who revealed himself in an individual event or in particular events" (VTT, 104–05).

It is true, of course, that claims of revelation (at least in the Christian sphere) entail claims of universality. But it is a mistake to assume that universality can be demonstrated by showing that the purported revelation is congruent with values known in advance to be universal. Rather, "universal validity is claimed for these revelations not because of their correspondence to some system of valid values previously discovered by men, but because they are revelations of the universal power and reality to which man and his values are required to conform." The problem with theology grounded in previously known values is that they "tend to make the revelations incidental or unnecessary and to reduce the individual religions with their individual founders to the status of examples" (VTT, 105).

The choice, then, is not (as it is sometimes claimed to be) one between beginning theology with a "dogmatic" confessional stance or beginning with an open and tolerant approach to values and then attempting to show the reasonability of the distinctively religious by undogmatically demonstrating the coherence of the religious with those values. This is misguided precisely because the values in question are derived from an ethics that dictates those values in the first place; "[a] theology which seeks to evade the dogmas of religion by founding itself upon ethics does not evade dogmatism, since every ethics rests on a dogmatic basis" (VTT, 105). So, if beginnings are to be made at all in theology, "dogma is unavoidable" (VTT, 105), and it is simply untrue to charge that confessional theology, in contrast to value-theology, is "dogmatic" simply because it begins with a priori values and then moves from those into theological construction.

The philosophical inadequacy with the value-theology Niebuhr considers lies in its "failure to take the principle of value relativity seriously."[6] Relativity of values, Niebuhr argues, does not signify that values must be considered relative to desires or consciousness, but at least that they are relative to the structure and process of the being for whom those values are values. To speak otherwise, as if values could be known and could be valuable apart from any knowledge of the observer for whom they are valuable, can be sustained only by means of "a vitiating abstractionism and the denial of the relative standpoint of the observer" (VTT, 106).

So despite the contributions of value-theology, Niebuhr argues that the final and necessary step of "complete abandonment of an approach from values known as absolute prior to the experience of God" must be taken if theology is to "become truly disinterested" (VTT, 110).

In this respect, Niebuhr observes, much can be said about the Barthian movement. He judges it to be correct both in what it asserts— "the absolute priority of God"—and in what it rejects—"the assumption that prior . . . to God's revelation of himself . . . the mind is in possession of a valid standard by means of which it can judge God and his revelation" (VTT, 111). While Barth may be right in these assertions and denials, he is misguided (according to Niebuhr) in thinking that theology can live by assertion and denial alone. Instead, the theologian must also demonstrate "where and how such denials and assertions have been made possible, and how they compare with the assertions and denials of other faiths" (VTT, 111). More general tools of philosophy of religion and other disciplines of religious studies become valuable here: "For this purpose a theory of religion, of revelation, and of religious ethics is necessary and such theories—today at all events—cannot be set forth without reference to valuation and values."[7]

In analyzing the role of value theory in theology, then, Niebuhr provides another philosophical support for the necessity of confessional method. Just as the importance of cultural and historical relativism is crucial in understanding the confessional nature of theological method, so does a close analysis of the role of value theory in ethics and theology also demonstrate that there is no alternative to beginning with "dogma" of one sort or another. This suggests that theology that is concerned to maintain its own independence and integrity must remain loyal to the particularities of its own revelation. Contrary to the mere dogmatic assertion of the particularities, however, Niebuhr can show that such an approach is warranted and valid from a general theory of religion, revelation, and ethics, which is grounded in a value theory that in principle is nonconfessional and open to general philosophical debate.

This attention to valuation has direct implications in Niebuhr's treatment of history and its relation to confessional theology. The possibility that religions are "stubbornly individual" is shown in the degree to which they form distinctive points of view because of the peculiar character of their histories. But the precise sense in which those histories are peculiar requires careful distinction. Histories of religions are not esoteric in the sense that the facts concerning the historical origins of religious communities and the founders or prophets of those communities are in principle unavailable to historians and others. As facts of history, they are necessarily the common property of all who possess the requisite historical tools and interests. And that is necessarily the case because they are facts of "outer" history, available to any external observer.

What distinguishes religious communities and forms their peculiar character is rather the valuational relation in which they stand toward those histories. The impersonal and objective data of the external history of a community is available to any historian, who is capable of valuing it as important or unimportant in terms of its influence on the areas of the historical concern or interest brought to that history. But a fundamentally different kind of relation obtains between the member of a religious community and the history of that community than that available to the "external" observer. Where the external historian is concerned to eliminate as much as possible the influence of the subjective and the partisan from the accounts of events in the historical life of a community, precisely the opposite is the case for the member of the community for whom history is necessarily "interested," concerned with the continuity of values and choices through time (MR, 44–54).

Niebuhr illustrates this distinction by citing two references to the U.S. Declaration of Independence. Lincoln's Gettysburg Address, which appeals to the values of the Revolution and applies those values to the situation at hand during the Civil War, represents an "inside" view of history—one that sees the continuities of value and moral commitment from the Revolutionary era into Lincoln's time. The *Cambridge Modern History* reference, on the other hand, merely narrates the events connected with the writing of the document and offers some general philosophical criticism of its linguistic vagueness (*MR*, 44–45).

Niebuhr summarizes these differences by noting that a member of the community views his own history from a fundamentally different perspective than that of the external observer. He states, "Though these various terms point to the same ultimate realities [they] are seen in different aspects and apprehended in different contexts" (*MR*, 45). External history is deliberately impersonal, concerned primarily with objects and with the "primary and secondary" qualities of events— those verifiable with reference to "permanent possibilities of sensation. Internal history is deliberately personal, finds its subject matter in subjects, and is concerned with "tertiary qualities"—with values (*MR*, 47–49).

This distinction illuminates a significant feature of human experience and of the formation of human communities, but it does not clarify their interrelationship. Niebuhr insists on the distinctness of the relations each type of history mediates between human individuals and communities, on one side, and the events of their histories, on the other. He also insists that the "terms the external historian employs are not more truly descriptive of the things-in-themselves than those the statesman uses and that the [historian's] understanding of what really happened is not more accurate than the [statesman's]" (*MR*, 45). But he will also indicate ways in which they have interconnections. While it is true that, "Being finite souls with finite minds in finite bodies men are confined to a double and partial knowledge" (*MR*, 61), Niebuhr is far removed from suggestions such as Holmer's that these points of view are wholly insulated from one another. It is important to note here his insistence on the importance of "external" views of the community's history for its "internal" history. Furthermore, the Christian community, the content of whose internal history testifies to a universal God, is especially obligated to strive to subsume all external history under the inner history of its grappling with the one God (*MR*, 61–66).

To summarize, Niebuhr offers two distinct although interrelated philosophical defenses of confessional theology. One originates in the general philosophical acceptance of relativism and is most fully developed in *The Meaning of Revelation*. The other emerges from a general philosophical theory of value. It stresses the relational character of valuation and the dependence of judgments of value on the nature of the entity doing the valuation. It attempts to show that all such value judgments involve an a priori choice of the "center of value" from which other value judgments are to be derived. Since the choice of this center is not itself subject to rational defense or derivation, revelation is essentially that of the value of a being which can be known to exist from other points of view, but whose value and proper role as the center of value cannot be so known. Because all such valuation is necessarily "confessional," this general philosophical theory of value constitutes a defense of the appropriateness of confessional Christian theology.

These two defenses converge in Niebuhr's "two-aspect" theory of history. They allow him to argue that the particular histories that form communities are necessarily relative (that is, not subject to general philosophical or theological defense as superior or normative for all cultures) and are distinguished from external views by the distinctive valuations placed on those histories by participants in the communities that result from them.

THE PROBLEM OF INTERNALISM

Confessional method in theology is most commonly assumed to entail "internalism." Essentially, this is the assertion that the claims and perspectives of Christianity (or indeed, of any particular religious world view) are "incorrigible," that is, not subject to criticism or reformulation on the basis of other spheres of knowledge or cultural perspectives. We have seen that this is indeed the claim of confessionalists such as Paul Holmer, Karl Barth and, to a lesser but still significant degree, D. Z. Phillips.

H. RICHARD NIEBUHR'S NONINTERNALIST PROPOSAL

Niebuhr's version of confessional method is to be distinguished from these positions on the question of internalism and therefore represents a significant alternative to the simple dichotomy between confessional positions identified with "Wittgensteinian fideism" or "Barthian positivism," and theological methods open in principle to the larger intellectual and cultural life surrounding them. Niebuhr's method represents a distinctive alternative among confessional positions; he "walks the tightrope" between legitimate confessionalism and fideism.

There may very well not exist a more uncompromising statement of confessional method than Niebuhr's: "As we begin with revelation only because we are forced to do so by our limited standpoint in history and faith so we can proceed only by stating in simple, confessional form what has happened to us in our community, how we came to believe, how we reason about things and what we see from our point of view" (MR, 29).

This resolute confessionalism applies at the "beginning" of theological construction, however, and not throughout its workings. Niebuhr insists that there are no universal starting points or points of view from which an apologetic can be constructed which will show on entirely value-neutral grounds that the particular story of the life of the Christian community is revelatory. He also insists that no set of values exists whose universality can be known a priori and with reference to which Christian theology can be shown to be justified. But he is far from asserting that the central claims of Christian faith can be formulated in a theological vocabulary or language that will be permanent and immune from revision with reference to the development of the surrounding culture.

In Niebuhr's own language, the problem of balancing the "highly conservative," confessional character of the "responsive, interpreting self" (RS, 96) and the need for "permanent revolution" as a part of the Christian's approach to reality (MR, 139) can be approached from two converging lines of thought.

One line is that which Niebuhr develops in *The Meaning of Revelation*, and focuses primarily on the interrelationship between "inner" and "outer" history (MR, esp. 59–66). The other is closely intertwined with the first, but uses a different vocabulary for formulating the problem and its resolution, giving preference to the language of "radical monotheism" and developing the implications of that language (*RM*, passim).

The arguments examined above depart from a general philosophical account of the historically conditioned character of human knowledge or of human valuation and prescind from the specific content of Christian theology. In these, however, Niebuhr attends to the very specific claim that Christian faith is normatively concerned with the confession of the "one beyond the many"—that is, with "radical monotheism."[8]

Niebuhr argues that human faith is rarely truly monotheistic, and is more commonly accurately characterized as either polytheistic (centering in many objects of devotion) or as henotheistic (displaying social faith in one object that is merely "one among many") (*RM*, 24–25). Since faith for Niebuhr simply refers to the inevitable reliance of a human being on a "center of value" and loyalty to the cause of that center of value for the meaning of one's life, actual lack of faith is a human impossibility.[9] But the more common forms of human faith are the reliance on a pantheon of "gods" (centers of value and of causes to which one may be loyal) or the exclusive devotion to one center of value which is not objectively ultimate but to which one gives exclusive loyalty (such as a nation or family).

Nominal Christianity may be subject to either of these inadequate understandings of faith. It may understand itself henotheistically as the religion of the "in-group" that constitutes the church or as a religion of Jesus Christ, or polytheistically as one center of value which is related to others such as a nation or family. In contrast, Niebuhr insists that Christianity is *normatively* not henotheistic or polytheistic, but radically monotheistic (*RM*, 60–63). Arguing that polytheism and henotheism show their inadequacy in human experience, Niebuhr insists that we are driven to radical monotheism. As such, our trust and loyalty are transposed to an object other than the particularity of either Jesus Christ or the religious community we call the church to a new reality. It is a reality that:

> We may call ... the nature of things, we may call ... fate, we may call reality. But by whatever name we call it, this law of things, this reality, this way things are, is something with which we all must reckon. .... This reality, this nature of things, abides when all else passes.

It is the source of all things and the end of all. It surrounds our life as the great abyss into which all things plunge and as the great source whence they all come. What it is we do not know save that it is and that it is the supreme reality with which we must reckon. (*RM*, 122)

Christianity and Christian thought are necessarily confessional in that they are grounded in particularity, "the concrete meeting with Jesus Christ" (*RM*, 124). Niebuhr argues that a "strange thing" happens within that highly particular history. Through the evolution of the understanding of the particularities of the faith of Israel and of the Christian church,

Our faith has been attached to that great void, to that enemy of all our causes, to that opponent of all our gods. The strange thing has happened that we have been enabled to say of this reality, this last power in which we live and move and have our being, "Though it slay us yet will we trust it." We have been allowed to attach our confidence to it, and put our reliance in it which is the one reality beyond all the many, which is the last power, the infinite source of all particular beings as well as their end. And insofar as our faith, our reliance for meaning and worth, has been attached to this source and enemy of all our gods, we have been enabled to call this reality God. (*RM*, 122–23)

The consequence of this claim about the normative content of Christian faith is that, while the foundation of the claims of Christian faith is, for Niebuhr, confessional, the faith that is mediated by that encounter "makes relative all those values which polytheism makes absolute." Although "faith is never so complete that it is not accompanied by self-defensiveness," the consequence of faith in the "one beyond the many" of radical monotheism is that we are involved in "a permanent revolution of the mind and of the heart, a continuous life which opens out infinitely into ever new possibilities" (*RM*, 126).

It is true that the origins and particularities of Christian faith derive from a particular history of the people of Israel and, even more centrally, from the history of the one man Jesus Christ. Because of this origin in historical particularity, Christianity is ineradicably confessional and continually must recur to that history. But the essential content that Niebuhr finds to have been revealed in the midst of that particularity is not itself particular. Rather, it represents a consistent drive in the direction of radical monotheism which, upon reaching fruition, undermines the henotheistic and polytheistic tendencies of such particularities in ordinary human experience. A two-edged sword, therefore, cuts both against specious universality that would seek to flee historic particularity and against a false and henotheistic confessionalism that is content to universalize its own particularities without subjecting them to the "permanent revolution" of criticism.

From this notion of faith as permanent revolution it follows that the attempt to fix past theological formulations as normative for all time or to claim for them the kind of immunity internalism does is a betrayal of the essential confession of the Christian that "God is acting in all actions upon you" (*RS*, 126). The God with whom the Christian confesses to deal is the one beyond the many, the ultimate source of all that experience—whatever its apparent source and however difficult it may be to subsume that new experience under the scheme of interpretation given by previous history of interpretation.

While the origin of Christian theology is "confessional" or historically particular, it cannot be charged with internalism precisely because of the distinctive content of that confessional claim. Rather than insulating theology from other disciplines and perspectives, Niebuhr claims that such radically monotheistic faith "opens the way to knowledge" by "remov[ing] the taboos which surround our intellectual life" (*RM*, 125). This is because radical monotheism alone is capable of facing the relativity of the relative. Radical monotheism alone is freed from the compulsions of henotheism or polytheism to ignore or flee the evidence of the relativity of the particular it has chosen as its "god" (*RM*, 125).

Similarly, in the constructions of value theory and ethics which derive from such radical monotheism, monotheistic theology is "enabled to proceed to the construction of many relative value systems," depending on the issue and the context with which it must deal. But "it is restrained from erecting any one of these into an absolute, or even from ordering it above the others" (*RM*, 112).

A monotheistically centered value theory is not only compatible with objective relativism in value analysis but *requires* it. Its fundamental dogma that none is absolute save God and that the absolutizing of anything finite is ruinous to the finite itself requires such relativism (RM, 112–13).

So we see that, from the perspective of radical monotheism, any absolutizing of the relative is ruled out. Niebuhr fails, however, to provide in this general cluster of arguments any positive advice regarding how theology is to make its particular choices. The danger of an all-relativizing position of radical monotheism is that it may undermine the necessity of choosing among relative systems one as relatively better than another. This problem is heightened when we remind ourselves of Niebuhr's insistence on the unavoidable biases that result from the fact that all thinkers begin from particular social and historical locations that color even the most fundamental levels of perception that they bring to new experience.

This point remains somewhat obscure even in the most sympathetic reading of Niebuhr; it is one with which he does not attempt to deal explicitly. This treatment is not cast in the language of radical monotheism, however, but rather in his discussion of the "inner" and "outer" history of communities.

Having distinguished generally between "inner" and "outer" history in Niebuhr's thought, it is now necessary to focus narrowly on what Niebuhr has to say about the interrelations between the two kinds of history. One could imagine a confessional position that drew distinctions similar to Niebuhr's and then proceeded to argue that there were simply no connections between the internal and the external history of communities, that is, that drew "internalist" conclusions from such distinctions. Niebuhr may not be crystal clear on the interconnections between these two kinds of history, but he does demonstrate that he does not believe that they are hermetically sealed off from one another.

Niebuhr points out that the earliest testimonies of the Christian church involved proclamation of a story, not demonstration of the metaphysical or philosophical cogency of a position. This occurred, according to Niebuhr, not simply because a story makes for a more vivid illustration than a sustained argument, but because "their story was not a parable which could be replaced by another; it was irreplaceable and untranslatable" (MR, 34).

The fact that the story was irreplacable raised the inevitable question of how it was related to the other story that was told of the community by those outside it. The solution of "so-called orthodoxy" was to insist that both systems of history operated on the same plane of experience, that of what is empirically available to the senses. Orthodoxy proceeded to insist that there were simply no connections between them. Reason, therefore, was left free to operate on the plane of the natural series of events, but was left aside when the supernatural sphere of revelation within that series was approached (*MR*, 55–56). This is a misguided solution, Niebuhr argues; it necessitates a dissolution of the unity of the self and of the community, postulating two spheres of experience and truth which have no connection—it is guilty of internalism! (*MR*, 57).

Niebuhr asserts:

> The two-aspect theory of history, like the two-aspect theory of body and mind, may be made necessary by the recognition that all knowing is conditioned by the point of view, that the exaltation of differences of understanding into differences of being raises more problems than it solves, that the intimate relations of subjective and objective truth require the rejection of every extreme dualim. But it is evident that the theory does not solve the problem of unity in duality and duality in unity. It only states the paradox in a new form and every paradox is the statement of a dilemma rather than an escape from it. (*MR*, 59–60)

In other words, Niebuhr recognizes much remains to be done in order to resolve the intellectual problem of the interrelations between the two aspects. Furthermore, he also recognizes that the formulation of a paradoxical relation is far from an intellectually satisfying resolution of the problem. Nevertheless, he insists that the clear statement of the paradoxical character of the problem is itself of intellectual value. This is the case because "it is important . . . that a paradox be correctly stated and that false simplicity be avoided. We have made some advance toward a correct statement of our dilemma . . . when we have recognized that the duality of the history in which there is revelation and of the history in which there is none is not the duality of different groups or communities, or when we have understood that this dualism runs right through Christian history itself" (*MR*, 60).

This statement of the paradox in no way provides a solution to the problem. It does make clear, however, that Niebuhr is far from holding an internalist position as evidenced by the mere fact that he can formulate his position on general philosophical grounds that are not subject to the charge of internalism. Instead, he offers his observations on the two-aspect theory of history as a part of a general theory about the features of distinctive communities which applies equally well to national groups, religious communities, and, by implication, to families. Unlike the proponents of confessional positions examined earlier, Niebuhr argues from a general theory of religion which shows on philosophically arguable grounds that all communities of human beings that are linked by shared values are necessarily formed around distinctive "inner" histories.

Merely saying that Niebuhr is committed to rejecting internalism in principle does not signify that he has an intellectually adequate method of dealing with the interrelationship between the two distinguishable kinds of history. In fact, he argues that there can be no speculative solution to the problem of the relation because there is "no continuous movement" between the two points of view. There is no "third" perspective that can include the others and provide the terms in which the perspectives of each can be expressed with equal accuracy; there is no way to circumvent the problem of incommensurable perspectives by direct access to "things in themselves" because "such knowledge of the nature of events is beyond the possibility of the finite point of view" (*MR*, 61).

While rejecting the possibility of a theoretical or speculative resolution of the dilemma, Niebuhr argues that there is a "practical" solution to the problem, a description of the "functional relationship" between the two perspectives (*MR*, 61–62). Even that description can only be offered confessionally, "as a statement of what we have found it necessary to do in the Christian community on the basis of the faith which is our starting point" (*MR*, 62).

Niebuhr claims that his suggestive remarks on the problem of this relation do not provide any theoretical resolution to the problem; nonetheless, they do suggest where the precise problem lies. They also provide the foundation on which a confessional theology that wishes to remain true to Niebuhr's fundamental insights and yet more adequate on this point might build a fuller epistemology.

The most striking remark Niebuhr makes on this issue, and one that most clearly distinguishes his position from internalist confessional positions, is the following: "Beginning with internal knowledge of the destiny of self and community, we have found it necessary in the Christian church to accept the external views of ourselves which others have set forth and to make these external histories events of spiritual significance" (*MR*, 62). Niebuhr cites the examples of the external critical views of Christianity offered by Celsus, Gibbon, Feuerbach, and Kautsky as examples of "external histories" of "spiritual significance" in the development of Christian self-understanding. He says of them,

> These have all been events in the internal history of Christianity. The church has had to respond to them. Though it knew that such stories were not *the* truth about it, it willingly or unwillingly, sooner or later, recognized *a* truth about it in each one. In so far as it apprehended these events in its history, these descriptions and criticisms of itself, with the aid of faith in the God of Jesus Christ it discerned God's judgment in them and made them occasions for active repentance. Such external histories have helped to keep the church from exalting itself as though its inner life rather than the God of that inner life were the center of its attention and the ground of its faith. They have reminded the church of the earthen nature of the vessel in which the treasure of faith existed. In this practical way external history has not been incompatible with inner life but directly contributory to it. (*MR*, 62–63)

From this observation alone it is clear that Niebuhr cannot be interpreted as advocating an internalist understanding of confessional method.

In addition to this line of argument, Niebuhr offers another that is essentially the same as that which we saw earlier in the discussion of radical monotheism: that even though the origin of the Christian community is necessarily historically particular in form, the content of the revelation so mediated is in principle universal—the revelation of the universal God. As he expresses it, "The standpoint of the Christian community is limited, being in history, faith and sin. But what is seen from this standpoint is unlimited. . . . Faith having apprehended the divine self in its own history, can and must look for the manifestation of the same self in all other events" (*MR*, 63).

It is because of this content that theology cannot, on Niebuhr's understanding of it, be understood internalistically. Even though it is necessarily the case that much of life and experience will remain obscure and uninterpreted in terms of the inner history of the community, the theologian is required by the nature of the object of his or her faith to attempt to discern in those events the pattern of meaning which coheres with the pattern derived from the particularity of the inner history of the community.[10] As Niebuhr expresses it, "faith having apprehended the divine self in its own history, can and must look for the manifestation of the same self in all other events" (*MR*, 63).

Niebuhr does not spell out in detail a method by which this ongoing application of the symbols and images of the inner history of the Christian community are to be applied to new historical experience. He does not intend confessional method to dispense the theologian from ongoing engagement (in principle, at least) with any aspect of human culture or historical experience.[11]

Niebuhr's understanding of theological method, then, shares certain features with other views that may also be classified as confessional in the broad sense of the term. Yet, this understanding differs on the question of internalism. Niebuhr shares with other defenders of confessional positions an insistence on the distinctive source and starting point of Christian theology. For reasons concerning both general epistemological theory and substantive understanding of theology, however, he parts company with them when they argue for the radically distinctive character of language-games that would immunize and insulate theology from events and intellectual movements which do not pertain (at least initially) to the inner life of Christian thought and community.

4 | AN OPEN CONFESSIONAL METHOD

V ERSIONS OF CONFESSIONALISM IN theology differ in their grounding, articulation, and ability to avoid simple internalism. Rejecting all confessional methods for the reason that all confessional method is internalist (in a pejorative sense) is incorrect. Some confessional analyses of theological method closely parallel constructions in other disciplines, especially scientific ones. This suggestive parallel between confessional method and contemporary epistemological debate offers a prima facie case for taking confessional method seriously as a viable approach to contemporary theology.

The task remains, however, to make these parallels explicit between confessional method and those epistemologies. The goal is to indicate where the more fully analyzed epistemology of the contemporary scientific view can be useful in extending and developing an understanding of confessional theology. Indeed, the analysis of the structure and operation of intellectual disciplines that has emerged in modern philosophy of science provides valuable tools and insights for two distinct but related tasks. On the one hand, descriptive analysis of the state of present-day theology is a worthwhile task in its own right, whenever it is heuristically helpful in bringing some intellectual coherence to thought about the state and nature of the field. On the other hand, the parallel assists in constructive theological work insofar as it clarifies and sharpens questions of method and boundary.

THE ANALOGY FROM SCIENCE

Superficially viewed, little seems to be gained by drawing the analogy between theology and a scientific discipline. Whereas a mature scientific discipline is highly organized both in its sociological structure and its intellectual organization, theology in the modern period is notoriously diffuse in both respects. Whereas a mature scientific discipline can point to assured results and even technological applications of its work, theologians are divided into competing schools which are barely able to comprehend each other's languages. Argument about fundamental questions of approach and methodology (a distinctive feature of undisciplined intellectual fields) consumes an ever-increasing portion of the activity of theologians. In this respect, too, theology contrasts with the practice of scientists within a well-established discipline who debate methodology only rarely, only during moments of disciplinary crisis.[1]

Yet, the approach to the analysis of intellectual disciplines best worked out in the context of philosophy of science can usefully be applied to the characterization of the broad range of differing theological methods, and to the more detailed analysis of an intellectually rigorous and precise confessionalism in particular. If the broad spectrum of contemporary theological methods is viewed from the perspective of the philosophy of science, then the theological scene clearly is neither wholly undisciplined nor structured along the lines typical of mature and fully disciplined fields. Rather, it is characterized by a large number of "schools," usually operating from the foundation of the philosophical or theological contribution of a single major intellectual figure, such as Barthians, Tillichians, Whiteheadians (Process theologians), Rahnerians. Within each school, terms and concepts are broadly shared. Often, there is considerable agreement regarding what claims and ideas can be taken as established and foundational to further inquiry and development of the school's thought. Often, at least, a sense of what major issues await treatment and analysis from the school's perspective is also broadly shared. Conversations between schools, however, are often characterized by incomprehension. Partly, the technical jargon characteristic of each approach is incomprehensible to others. Additionally, however, it also reflects a divergence of opinion regarding the nature of the questions thought to be pressing in the school, with one regarding as settled what from another perspective is wholly unresolved or uninteresting.

Such division of a subject area is typical of what Masterman called "multiple-paradigm" fields. Local progress that reflects the shared activities of a group of researchers is possible. Such progress is recognized only by the members of the subcommunity of researchers who are committed to that particular method or approach and often considered mistaken or wholly beside the point from the perspectives of other communities of scholars.

Even in theology, however, this state of affairs is not a necessary one. Some historical periods have been governed, like that of a normal science, by a single paradigm. During such periods, theologians work within a broadly shared framework which, although surely looser than the paradigm that governs a mature natural science, shares many features with such a discipline. During those periods, many theologians take certain approaches to inquiry for granted, accept some classic works of their field as assured results, and share a common view of the problems that remain to be resolved. Certainly Thomism functioned in this manner until recently in Roman Catholic theology, and Protestant scholasticism managed a similar clarity of purpose and method for a substantial period of time. Even in modern theology, neoorthodoxy dominated during the first half of the twentieth century in Europe and the middle third of the century in North America; biblical theology enjoyed a briefer dominance in the United States during the 1950s—these are examples of coherence that large numbers of scholars shared.[2]

Yet in the modern context, these periods seem to be like those Masterman identifies in which normal science set in prematurely, before the governing paradigm was so developed and articulated that it could sustain successful research programs over a period of generations.[3]

Of course, application to theology of the kind of analysis of intellectual fields that has been developed in the philosophy of science could be extended indefinitely. It is useful in describing many of the disciplinary characteristics and difficulties of theology and other similarly diffuse fields. The main use of this perspective, however, is to be found in the implications of the parallels between the fundamental epistemological claims of scientific disciplines and the particular theological method advocated by careful confessional theological thinkers. The main function of these parallels concerns the ways in which these claims may be helpful in constructively extending confessional method.

Confessional theology in general can also be seen as merely a school such as those discussed above; on this reading of things, a confessionalism along Niebuhrian lines is only a "school within a school." The utility of the material gleaned from the philosophy of science could then be seen merely as providing a classification scheme that allows us to locate confessional theology on a taxonomic chart of the multiple-paradigm field called theology.

The preceding chapters show, however, that the most nuanced among confessional positions are not subject to the usual criticisms and dismissals of that method and provide a viable method for contemporary theology. To that degree, then, not only a descriptive and taxonomic judgment regarding the classification of an approach to confessional theology along Niebuhrian lines but also a normative one was offered.

The superiority of an understanding of confessionalism along Niebuhrian lines in comparison to nonconfessional positions has been argued so far only by implication in the exposition of Niebuhr's own defense of his method. It is now possible to draw on the "emerging consensus" regarding epistemology which has developed in the philosophy of science over the course of this century to offer a more explicit defense of Niebuhr's fundamental insights and of the essential confessional claim regarding "starting points" in theology.

The fundamental claim of confessional method is that theology must begin with revelation, with "what we see from our point of view."[4] The attempt to do otherwise is the attempt to find the kind of nonparticularistic starting point and neutral foundation that social science and history have shown to be an illusory goal. This claim, for Niebuhr, was grounded primarily in the observation that relativism had come to be an axiom of social and historical research, and originated from his close engagement with the work of Ernst Troeltsch from the days of his dissertation on.[5] In view of the developments in philosophical epistemology associated with the philosophy of science since Niebuhr's time, however, it is now possible to make a stronger case in favor of this fundamental claim of confessional method—one that draws on a wider range of intellectual disciplines.

The inescapability of relativism in all thought claimed by confessional method closely resembles that claimed by contemporary philosophers of science. As we have seen, they criticize Logical Empiricist efforts to provide a nonparticularistic and a priori distinction between the scientific and the nonscientific which does not depend in any way on the actual state or historical development of science. They have been able to demonstrate that the quest for such a distinction, and for an a priori definition of what will count as "scientific" in a particular discipline, is misguided and fails to account for what practicing scientists actually do. Furthermore, they have also been able to show that scientific rationality itself is not historically uniform, and that what counts as a scientific question or as a scientific datum varies even within a discipline, depending on the governing paradigm at a given stage of its development.

Much of the criticism of confessional theology from other theological methods is centered on its claim about the necessity of a "starting point" that is particularistic. Such starting points are not demonstrably valid on grounds of rationality equally available to "outsiders," to the religious community as well as to insiders, such criticism points out.[6]

One conclusion to be drawn from the developments in philosophical epistemology during this century is that, at least in its most basic form, this criticism is misguided. It asks for a theology to meet an epistemological standard that is unattainable in principle. If even disciplines that are unquestionably paradigm cases of disciplined inquiry and knowledge (such as physics or astronomy) cannot sustain any such distinctions, surely theology cannot either.

Contemporary epistemology shifts the "burden of proof" away from confessional method's emphasis on the indemonstrability of the validity of its starting point and onto the epistemological assumptions of the allegedly nonconfessional positions.[7] To this extent, then, contemporary philosophy of science provides a valuable resource for further defending Niebuhr's claims (shared broadly by all confessional positions) on the question of the "starting point" of theology. It therefore is a very valuable supplement to the kinds of arguments from the sociology of knowledge Niebuhr uses to support the two-aspect theory of history that is at the foundation of his inner/outer history distinction.

The quest for foundations for scientific inquiry which are theory-neutral, the demand for the wholly disinterested collection of all data which presents itself to the observer, and the incorporation of all such data into the structure of the existing theory—all these purported requirements of knowledge—are prescriptions that do not pertain to the working of science or any other mode of systematic thought. Indeed, such a description of research is approximated only in undisciplined inquiry that has not yet determined which lines of inquiry are likely to extend the existing model of the discipline in question in fruitful ways. At that stage, perhaps, it is still collecting information in an unsystematic and unfocused manner. But that relatively unorganized collection of information ceases when even a glimmer of a paradigm has established itself among a group of researchers. From that point on, future observation is guided and data sought in accord with the indications of that paradigm.

From these observations a number of conclusions follow regarding the relationship between epistemological inquiry and contemporary scientific method. The philosophers of science examined here are unified in their insistence that disciplined inquiry on the basis of assumptions and methods shared by a group of researchers does not emerge from discussions of method per se. Rather, perennial discussions of method characterize a field which has not yet moved beyond the stage of the "philosophy of" the subject and which has not emerged into disciplined inquiry.[8]

Furthermore, such a period of methodological discussion is not brought to an end and substantive work begun when there is a definitive conclusion to the discussions of method. Rather, disciplined thought is possible only when a person or group embraces one of the competing methods of the predisciplinary period and is able to demonstrate results that win general acclaim and acceptance. In other words, one clear implication of the epistemological literature is that breakthroughs in methodology never result from discussions of methodology. It is only when substantive results achieve acceptance throughout the group of researchers and acquire the status of "assured results" that a field escapes the morass of interminable debates over method.

The attempt of the Logical Empiricist philosophers of science to determine a priori criteria for the definition of a scientific discipline is one which, whatever the disagreements regarding detail, the consensus of historicist philosophers of science judges to have been a failure. To this degree, therefore, the confessional claim that theology cannot resolve the methodological questions of "starting point" and procedure in the abstract finds strong support from the developments in the philosophy of science and philosophical epistemology. Niebuhr's assertion that "the concept of nature has a history and . . . men respond to natural events in varying ways in different periods of social history, on the basis of their different interpretations"[9] foreshadows and anticipates the conclusions drawn from the analysis of the working of the sciences. To this degree the insistence on a confessional starting point is generalized to the claim that any orderly inquiry must have a similarly non-a priori point of departure.

The constructive implication of this agreement is that the tendency of much of contemporary theological debate to focus almost exclusively on questions of method requires close and critical analysis.[10] Such preoccupation is most easily understood if it is taken as a symptom of the chaotic state of contemporary theological debate and of its inability to achieve anything resembling assured results that win general recognition within the field.[11]

However, if the intent of such methodological debates is to achieve through methodological clarification alone the agreement that will make possible progress toward substantive progress, they are essentially pointless tasks: they invert the relationship between actual substantive progress in a field and methodological clarity. Methodological clarity follows rather than precedes substantial disciplinary progress.

The other issue that confessional method raises (although it is hardly unique to it) is that of the "boundary problem." This refers to the method by which a confessional position determines whether perspectives, information, and criticisms that arise "outside" that position are to be incorporated into the "inner" perspective of that confessional tradition.

As seen earlier, many forms of confessional theology claim to avoid this problem entirely. Their usual strategy is to claim the immunity of uniquely religious discourse from contact with, and therefore criticism from, other perspectives and conceptual frameworks. Barth seems to represent an alternative approach, claiming (without example) that all other perspectives can in principle be subsumed under the theological. Despite differences in detail, such claims are characteristic of internalist confessional positions.

Such positions are subject to numerous difficulties, unable to account for the actual working of theological discourse within that framework. They also fail to provide a useful set of categories within which to account for change in that discourse over time. Only with great difficulty, if at all, can they explain the demonstrable changes in theological conceptual frameworks over history in response to changes in the philosophical, scientific, and social conditions in the culture. The necessity of recognizing that such changes do, in fact, occur results in inconsistencies in confessional positions that attempt to articulate a coherent internalist position and lead to the understandable suspicion regarding the intellectual power and adequacy of confessional method in general.

Not all confessional positions are equally internalist, however. Niebuhr's and Fiorenza's positions explicitly acknowledge both the reality and the positive value of such interactions between theology and other cultural perspectives, and the necessity for theology to incorporate some of those "external" perspectives into itself. In Niebuhr's case in particular, insistence on this necessity does not derive from the mere unavoidability of such incorporation (which might be accepted reluctantly and with regret). Rather, it is generated from two normative theological claims: (1) that the hearing and accepting of external criticisms is an essential control on the tendency of an unchecked confessionalism to become uncritical of itself, and, (2) that the need to incorporate new information and perspectives results from the normative claim of truly monotheistic Christian theology that the object of its inquiry, the God it confesses, is necessarily the "one beyond the many," encompassing (in principle) all of reality and knowledge.

But, while these acknowledgments on Niebuhr's part make his understanding of confessional method superior to others on both descriptive and normative grounds, his own articulation of his view left a major unclarity to be resolved. This problem is how a confessional method determines which initially external perspectives carry valuable criticisms or perspectives the inner history requires, and which are to be appropriately rejected on grounds of essential and irresolvable incompatibility with the confessional perspective. This is a serious and genuine problem for the extension of a theological method that intends to remain loyal to Niebuhr's essential insights. The actual doing of theology requires constant judgments regarding precisely such questions. Yet all the examples Niebuhr gives (such as the incorporation of Feuerbach's or Marx's criticisms of religion into the inner account of Christian theology [*MR*, 62–63])[12] are the judgments of hindsight, not helpful guides to a theologian attempting to make such judgments regarding contemporaneous external perspectives.

This problem is closely analogous to that of a thinker working within the framework of a particular understanding of a scientific discipline and attempting to decide whether a new paradigm should be adopted or whether data that vary from that predicted by the dominant theoretical framework are significant and require revision of that theory. This set of issues has been discussed at great length in the philosophy of science literature. It centers there on questions of the rationality of the acceptance of paradigms, of paradigm shifts, and of the continuity and discontinuity through change of disciplinary perspectives.

The question now is whether the analog between the issues in their respective settings is close enough that the philosophy of science discussion can be useful in clarifying and, possibly, extending confessional theological thought on this issue. If help were forthcoming on this problem, one of the major gray areas of even the best confessional positions would be substantially mitigated.

To begin such an analysis one must note that the use of the philosophy of science material in this discussion applies at a fundamentally different level than it did in the earlier discussion of the taxonomy of the many kinds of thought which are broadly characterized as "theology." These categories were applied earlier to theology to show that at best it can be seen as a multidisciplinary field divided into schools and, at worst, as wholly undisciplined thought. At this juncture of the discussion, confessionalism will be treated exclusively as a method and in its most nuanced version, as if it were a discipline in its own right.

Such a treatment presumes the value of attempting to work from within a fundamentally confessional framework to extend and develop constructive confessional theology. For that purpose an attempt must be made to determine the degree to which it can be seen as analogous to a mature scientific discipline setting out to make judgments regarding the incorporation or exclusion of novel theories, paradigms, and kinds of data. The goal is to determine whether the insights that derive from the close study of the working of scientific disciplines in the philosophy of science can prove helpful in that extension of confessional method's clarity about what has been referred to here as the "boundary problem."

Mature scientific disciplines are routinely confronted with suggestions that greater attention be paid to data and observations that seem inexplicable or irrelevant. In addition, they confront the claim that an alternative model of the fundamental explanatory ideals of the discipline should supplant the existing one. One of the characteristics of a mature and disciplined science, and one of the reasons for its success in mobilizing a large group of workers in attacking a single problem without central institutionalized organization, is precisely its ability to ignore such suggestions during periods of normal science. In such cases, it is successful in devoting its attention rather single-mindedly to the elaboration of the regnant model rather than perpetually challenging and reassessing its foundations.

The question in a scientific discipline is how to determine when it is necessary to suspend this normal procedure of working within the bounds of the established paradigm to attend to such challenges. More fundamentally, one would like clarity regarding the criteria that might be able to guide the member of the discipline attempting to make that choice at a particular juncture in the evolution of the field. But here is precisely the point at which abstract theory fails: It is agreed among the historicist philosophers of science that there are no a priori criteria to which a researcher can appeal to resolve such issues. Further, historicist philosophy of science notes that there are inescapably nonrational factors operating in the acceptance or rejection of new perspectives. As Kuhn writes, "Paradigm change cannot be justified by proof."[13]

Kuhn is often accused of implying that the maintenance of an old or the acceptance of a new paradigm is a wholly irrational and even random matter. But this is not the whole story. As seen earlier, rather than being amenable to an a priori definition of the criteria, it is more accurate to describe the development of a scientific discipline over time through the changes in paradigm which guide it in evolutionary terms. As both Toulmin and Kuhn insist, the best analog to the development of a discipline is the progressively better adaptation of a species to its environment. This process occurs without the benefit of a preestablished telos. It results from the interaction of the species and the environment that through natural but undirected processes produces ever more elaborate specialization and articulation.

Similarly, scientific disciplines become better adapted, broader and more effective in their accounting for the world. Like biological evolution, however, this process cannot be viewed as approaching asymptotically the ideal limit of a fixed structure of nature which, ideally, it will perfectly mirror.

For this analogy to be descriptively helpful, however, it is necessary to locate the feature of the development of scientific disciplines which corresponds to the source of evolutionary pressure, that is, to natural selection. This analog is what Toulmin calls the intellectual ideals of the discipline in question.[14] This is the common conception of the "general forms . . . a complete account of [the area of inquiry of the discipline] should take" (*HU*, 151). Such intellectual ideals specify with a precision proportionate to the degree to which the discipline in question is a mature science the "intellectual shortcomings" (*HU*, 152) of the discipline. The more highly disciplined the field of inquiry is, the more precisely it is able to agree on the outstanding problems of the discipline and the significance and weight of those outstanding problems for the regnant paradigm. The ideal provides the "template" that is brought to the phenomena the discipline has set itself to describe. The misfits between the template and the phenomena deemed relevant by members of the discipline provide its "characteristic problems," its research agenda.

The striking feature of highly disciplined fields such as the natural sciences is the precision of their intellectual ideals, the detail with which the template guiding their inquiry predicts the phenomena they will observe. Directly in proportion to the precision of the template, the researcher is able to identify specific anomalies between the predictions of the template and the actual phenomena. These anomalies are the counterpart to natural selection in the evolutionary analogy. They provide the "evolutionary pressure" that drives the discipline to pursue such anomalies as pressing research problems and, after extended failure to resolve such problems, to begin to cast about for alternative models that might allow progress to be made in resolving them.

As we saw earlier, it is precisely at such junctures that there are no general guidelines to help the research community decide whether the proper strategy is to continue to work at resolving such problems through research guided by the old paradigm, or whether the time has come for the shift to an alternative model in hopes that it will provide a means for making further progress past this impasse. It is at such points of disciplinary confusion, defined by the failure of the present explanatory ideal to resolve problems that the discipline agrees to be pressing, that dual or multiple paradigms emerge until one of them manages decisively to solve the puzzle set by the impasse.

The question of the relevance of this analysis of highly disciplined fields for the constructive extension of confessional theology, therefore, turns on the question of the nature and precision of the intellectual ideal that such theology brings to experience. This issue, in turn, depends on the extent to which it can agree on the precise range of phenomena it purports to interpret, and the degree to which the assumptions embedded in the essential features of the confessional position are analogous to the replaceable paradigm of a scientific discipline.

It is obvious at the outset that the anomalies in the case of theology are necessarily much looser and less focused than in the sciences. The precision with which physics, for example, is able to predict a particular measurement derives from a mathematical modeling of expectation which can have no parallel in theology. And yet, there is a looser but nonvacuous sense in which theological constructions do generate expectations of the world and of experience, and those are expectations which can be disconfirmed. Our goal, therefore, is to specify as precisely as possible the nature and functioning of such models and the significant similarities and differences between them and the templates of scientific disciplines.

Perhaps the closest analog to the intellectual ideal of a scientific discipline in confessional theology is the awareness that even though the core confession of Christian faith has its origins in a highly particularistic and irreplaceable story of its own inner history, the "divine self" encountered in that history is confessed to be identical with ultimate reality. This conviction grounds the further claim that Christian faith must "look for the manifestation of the same self in all other events" (*MR*, 63).

The formulation of this ideal with reference to action is that found in Niebuhr's characterization of the model of the responsible self: "God is acting in all actions upon you. So respond to all actions upon you as to respond to his action" (*RS*, 126).

In these formulations we find the intellectual template that guides the work of constructive confessional theology: that all events are (at least ideally) to be brought into an interpretive framework that presumes that they are manifestations of the same divine self whose nature has been made known in the inner history of the confessional community. The task of theology, then, is to extend and apply this paradigm to the new phenomena of the ongoing historical experience of the community. Phenomena that initially appear to fall outside the explanatory power of this model, that cannot be readily assimilated into the historically developed and articulated self-understanding of the community, create the "research problems" of theology. The explanatory goal is to seek to subsume those experiences under the explanatory ideal of finding the action of the same God confessed by the community in the midst of those new experiences.[15]

Because of the radically monotheistic character of the faith normatively confessed by Christians, the range of phenomena to which theology is relevant is nothing less than the whole of reality. While there is a sense in which this is true of scientific disciplines as well, theology is distinguished from them in its relative inability to specify a standard set of more proximate goals. That is, while it may be the goal of mature sciences in some sense to be able to account for the whole range of phenomena given in experience (as in sociobiology's attempt to subsume anthropology, social science, and even psychology under biology; or the attempt of brain physiology to account for psychological events), scientific disciplines are better able than theology to discriminate areas of experience which are relatively closer and relatively more remote from their disciplinary concerns as the discipline is conceived at any given time. They can use that discrimination as a means of assigning relative priorities of urgency to phenomena awaiting investigation. Theology, on the other hand, precisely because of the global character of its concerns and interests, its concern with "limit questions" that transcend the more focused and specified questions of a mature scientific discipline,[16] has substantially greater difficulty determining which kinds of information and perspectives are immediately germane to its inquiry.

This is partially true. There are kinds of information and data that all theologians consider relevant to their work, such as the interpretation of the Christian Scriptures, the history of Christian dogma and creed. But the distinction between theology and scientific disciplines is still significant when one considers that there is virtually no agreement on the nature and degree of the relevance of such considerations (for example, is all theology essentially determined by biblical interpretation, or only informed by it?) or, indeed, on how they should be understood and interpreted.

Furthermore, once one departs from such obviously theological or religious considerations as Scripture and church history and turns to the question of the relevance of other disciplinary perspectives (such as literary critical theory or the results of inquiry in particular natural sciences), one encounters only diversity. Different theologians plead for the value, or even the crucial significance, of perspectives that other theologians fail to see as central to the theological task—or even as relevant or significant.

So a major point of contrast between mature scientific disciplines and theology is that a scientific discipline, because of the much more precise nature of its governing paradigm, is able to discriminate between kinds of information and perspectives that are immediately germane to the field (even if, for the present, they cannot be accounted for in terms of the regnant paradigm). Theology, because of the diffuse but all-encompassing nature of its explanatory ideals, is necessarily incapable of making uniform determinations of the kinds of exclusions of perspectives and information appropriate to its work at any given time.[17]

Such observations go to show that, while the analogies between the fundamental epistemological claims of historicist philosophy of science and a sufficiently nuanced and subtle confessional method in theology are significant and helpful in many ways in supporting and supplementing confessional method, there are also extremely important differences between theology and a mature scientific discipline which restrict the range of applicability of those analogies.

Again, one reason for the relatively rapid and shared progress of a scientific discipline is that ability of such fields to focus attention narrowly on a range of phenomena which the governing paradigm of the discipline indicates to be appropriate and, correlatively, to exclude many other kinds of information which the relative precision of the model allows it to exclude. Theology lacks an "exclusion principle" of even approximately equivalent precision. Hence, it is much less able to narrow its concentration to a precise and communally identified range of phenomena awaiting description.

This difference is significant for the "boundary problem." Even though in the sciences it has become clear that there are no a priori ways of determining when new perspectives and paradigms should be accepted, the ability to narrow the range of questions that are on the "agenda" of a scientific discipline allows broad disciplinary agreement regarding the pressing explanatory problems of the discipline. Since theology is so much more diffuse, however, it is not possible to define clearly those problems and perspectives that, while "outside" the confessional tradition as it stands at any given time, deserve to be included in the "inner history" of that tradition. It is also (in contrast to the sciences) not possible to determine which perspectives, phenomena, and methodologies require priority consideration from the perspective of the understanding of the field current at that time (that is, what pressing phenomena cannot be adequately explained by the present paradigm even though they are clearly within the proper range of phenomena for the discipline?).

On the other hand, our analysis of the working of science in periods of insecurity regarding the adequacy of its present paradigm suggests the inappropriateness of asking for precision regarding the boundary problem as a precondition for accepting the fundamental adequacy of confessional method. It would be asking a diffuse field such as theology a methodological formula that even highly precise and focused disciplines cannot in principle offer. To the extent, then, that the "boundary problem" is subject to inherent irrationality, even in the most mature of scientific disciplines, theology can hardly be expected to do better. As in the sciences, the proof of the correctness of opting for a new paradigm or of attending to new data can only be given after the fact. Only when a model has shown itself to be successful in explaining more of experience and of advancing the collective knowledge of the relevant community has the method shown itself to be "correct" in the only relevant sense: successful. Similarly, in theology, the appropriateness of incorporating an external perspective into the inner history of the confessional community can be demonstrated only by advancing the range of experience that can be brought under the rubric of its understanding of God acting upon it. If it is true of the natural sciences, it is surely even more true of theology that "as in political revolutions, so in paradigm choice—there is no standard higher than the assent of the relevant community" (SSR, 94).

As in the sciences, the goal of confessional theology may be to explain the whole of experience by the extrapolation of its confessional perspective to cover new ranges of experience and data. But equally as in the sciences, the path of knowledge lies in the judicious determination of thinkers within the field regarding when to continue to work exclusively within the existing framework, attempting to cope with and explain new ranges of experience in its terms, and when to decide that the existing framework must be recast, attending to new data or incorporating new paradigms. If it is true of the sciences that there is no standard higher than the assent of the relevant community to guide such decisions, more can hardly be asked of theological method.[18]

One last area of significant comparison and contrast between confessional theology and scientific disciplines remains: that of the status and replaceability of paradigms. This is significant because historicist philosophy of science, like confessional method in theology, stresses the role of the structuring assumptions that guide thought and research within their frameworks. But philosophy of science also stresses the fact of the alteration of those assumptions as old frameworks are abandoned, concepts altered, and research under the aegis of new intellectual ideals begun. In other words, despite disputes regarding their radicality and frequency, historicist philosophers of science are united in their claim that scientific disciplines experience periods of discontinuous change in fundamental disciplinary assumptions. Are there the equivalents in theological discourse—"theological revolutions"?

On the one hand, all confessional thinkers stress the necessity of a confessional starting point in theology, a starting point that is an "irreplaceable and untranslatable story" (*MR*, 34). The exposition of the process by which theology advances and adapts stipulates an unchanging assumption brought to all experience, which defines the essence of understanding of Christian faith: that all experience is a manifestation of the working of the God known in the confessional tradition. From that assumption it follows that all experience is, at least in principle, susceptible to interpretation in terms of the working of that God. The abandonment of this assumption would constitute the abandonment of the Christian theological endeavor.

On the other hand, this concern with unchanging assumptions is balanced by an emphasis on the relativity of perspectives of various religious communities and historical periods and the awareness that theology has, in fact, changed over time. The historical record is clear that theology has incorporated into itself perspectives initially external to the religious schema of the community. Accounting for these facts requires a recognition that much of the explication of Christian theology does, in fact, change in response to changing cultures and historical conditions.

The fact of conceptual change in theology is too obvious to deny. Theology adapts as it strives to explain new and initially unfamiliar experiences in terms of its confidence that it encounters the God of its confessional tradition in all experience. Sometimes these adaptations are the theological equivalent to paradigm shifts. They can involve the adoption of new explanatory frameworks into the theology, often frameworks borrowed from other, initially alien, disciplines. Sometimes adaptations require the alteration or abandonment of the assumptions that govern theological inquiry at particular points in its development. It is precisely to this feature of theological development that Niebuhr intends to draw attention in characterizing radically monotheistic theology as "permanent revolution of the mind and of the heart, a continuous life which opens out infinitely into ever new possibilities" (*MR*, 34).

On the other hand, confessional theology insists on a starting point of theology in the constant and irreplaceable story. How can this claim be harmonized with the idea of permanent revolution? It is important to note that the story must be not only irreplaceable, but also "untranslatable" (*MR*, 55–56). This locates the inescapably confessional aspect of theology not in particular conceptual formulations of the theological tradition at a particular point in its development (in its "translations"), but in the pretheoretical images and "stories" that inform a way of seeing the world of experience as a manifestation of the divine purpose. The attempt to "spell out" the connections between the irreplaceable stories and images of the confessional aspect of tradition and the fully articulated and conceptualized world of theology proper is a constantly changing and developing process. But it is a process that is grounded in the images and stories which come from the confessional aspects of the tradition. These are the point to which the tradition must continually recur to ensure that its "translations," that is, particular conceptually articulated understandings of theology, are harmonious with those core image-guided perceptions.[19]

One can abandon those central images and stories, in principle, and simply cease to believe that there is any point to attempting to see all actions upon one as actions of the same Divine Self as the confessional tradition has led one to expect. That would not be a paradigm shift in theology, however. Rather, it would be a shift that causes one to abandon, if not the entire theological enterprise, at least the enterprise of Christian theology. Religions do die; there is no reason to think the demise of Christianity inconceivable in principle.

There are various levels of conceptual articulation of the content of confessional belief. Some are highly theory-specific ones (such as Holmer's example of transubstantiation). Others refer to broader beliefs articulated in a large number of theoretical and historical contexts ("real presence" generally, or the claim that Christ is redemptive). And most broad of all, there are those assumptions that define Christian religious belief in any form (for Niebuhr, for example, beliefs such as those associated with radical monotheism).

With those various levels in mind, one can see that an adequate understanding of confessional theological method implicitly acknowledges and even embraces change, even fairly fundamental paradigm change. At the level of the conceptual articulations appropriate to Christian theology at various points in its historical development, such changes are necessary and important. It is important to note, however, that there are levels at which the "story" is "irreplaceable" and theology necessarily confessional. This is not because one cannot imagine ceasing to find that story to be revelatory, but because the loss of that belief would necessarily imply ceasing to be religious, at least "Christianly" religious.

Confessional positions, for which the whole of theology is claimed to be unchangeable, seem underdeveloped. A more subtle distinction between the various levels of theological discourse is required. There are fundamental confessional elements that form the perception a believer brings to experience and that are not justifiable on grounds of general rationality. They are irreplaceable short of ceasing to find value or meaning in a particular religious orientation at all.[20] But there are also theory-dependent conceptual articulations of belief which are highly variable and malleable in the light of changing experience.

This more nuanced version of a confessional position closely parallels a historicist philosophy of science. While a scientific discipline may adopt new theoretical frameworks at various points in its development, it too has higher level assumptions that govern the discipline even in the midst of change. Those assumptions cannot be abandoned, unless the discipline ceases to be. Physics, for example, may be able to shift from a Newtonian to an Einsteinian reference frame as its fundamental theoretical structure at a given point in its development. But it cannot give up the higher level assumption that physical reality is amenable to discovery and description by rational means. This is not because it is a "given" of nature that such discovery is always and indefinitely possible, but because abandoning that assumption would cause the field to cease to be physics altogether.

There are also no a priori methods for determining whether and when abandoning of a discipline should occur at all. Just as physicists argue within a framework set by fundamental assumptions regarding nature, so do astrologers. As physicists cannot cease to hold the fundamental assumptions of physics without ceasing to be physicists, neither can astrologers abandon their fundamental assumptions (for example, that individual human life experience is decisively affected by the positions of celestial bodies). Disputes will occur among astrologers, but at a level "lower" than the shared assumption that a proper horoscope is determinable for a given individual.

If a person ceases to be an astrologer, it will not be because the astrological perspective has been directly confronted and defeated on its own terms. Rather, it will be because another perspective seems to hold a greater "promise of truth"[21] than did the previous one for that person. From the latter perspective, not only individual claims made by astrology, but the point of the whole astrological enterprise will no longer engage imagination and thought.

The conditions necessary for an individual to become disenchanted with astrological assumptions cannot be specified in advance. One always has the option of ignoring disconforming evidence. There are always ways of immunizing the theory from such disconfirmation rather than abandoning the governing assumptions of the framework. Human ingenuity is nearly infinite in its capacity to maintain basic beliefs if there is a strong urge to do so.

As seen in the discussion of the philosophy of science, the test of the value of the governing assumptions of a particular framework cannot be demonstrated a priori. The ability of those working within that framework to produce results accepted by the relevant community serves to "confirm" the assumed framework. Disciplinability is attained in a nascent science not by resolving the a priori question of the nature of science per se, but by showing that the adoption of a particular paradigm actually is able to produce communally accepted results.

Similarly "dogmas is unavoidable"[22] at the foundation of theology. It is justified by its success in describing experience accurately, as that is determined by the verification of "social companions" (RS, 80), fellow believers who share the perspective of "what Christians see from their limited point of view in history and faith" (MR, 31).

The plausibility of the foundation of any field of inquiry is determined wholly by the existence of a community that continues to believe that the intellectual ideal governing the inquiry of the community holds a "promise of truth." As long as a community confesses to have encountered God in the particularities of the inner history of the Christian confession, there will be a community that "seeks to interpret each new occasion by assimilating it to an old encounter" (RS, 96).

No community can ensure its survival through an a priori defense of its existence. That realization is an essential part of the Christian confession. Any alternative would reflect "defensiveness" and would distract the community from its proper work of recalling the story of Christian life and analyzing what Christians see from their limited point of view in history.

This raises a profound, unresolved dilemma in confessional theology: Where is such a community to be found in the modern world, and what is its nature?

Some descendants of Niebuhr's fundamental perspective stress the necessity of a community sufficiently cohesive to serve as the locus of such mutual criticism within a common framework. Many of them have been led by such thoughts increasingly to stress the need for an essentially sectarian church. Evident in the work of Stanley Hauerwas is Niebuhr's emphasis on the essentially "inner historical" character of the story that forms the core of Christian identity, which is taken to move one toward a relatively self-contained community that witnesses by virtue of its "differentness" to others.

On the other hand, Gordon Kaufman and James Gustafson, equally legitimate claimants to the Niebuhrian tradition in the contemporary context, stress the all-relativizing character of Niebuhr's radical monotheism, and go farther than he did in emphasizing the highly limited epistemological situation of confessional traditions.[23]

The task for constructive application of confessional method is to offer substantive theological proposals to communities of discourse. The boundaries of the relevant community can no more be specified in advance than can the foundatins of inquiry. Indeed, communities are formed and reformed in a complex interactive dance between intellectual construction and sociologically determined boundaries. Some communities find new proposals generated from within their midst to be so powerful that they reorder and restructure the boundaries of community. At times communities find the intellectual articulation of belief in a neighboring community so powerful that they adopt it as their own (witness the appropriation of Paul Tillich's theology by some Roman Catholics).

Confessional theology is fundamentally correct when it points to the interaction between the life of particular religious communities—their symbols, their rituals, their forming of individual minds and hearts—and the more abstract formulations of theology. While it is the besetting temptation of confessional positions, it is wrong to think that that relationship is static. It is wrong epistemologically to imagine that religious language is sharply and definitively to be distinguished from other language. It is wrong religiously to imagine the religious community to be *in principle* immunized from the vicissitudes of historic encounter with other communities and their ideas, and ultimately with the God who acts in all actions upon it.

CONCLUSION

The highly developed analysis of epistemology emerging in contemporary philosophy of science is useful in clarifying the state of contemporary theological diversity. The model of a disciplined body of inquiry that has emerged through the analysis of the working of scientific disciplines is both analogous to Niebuhr's fundamental epistemological claims and useful in supplementing the arguments he offers in favor of that method. Besides offering further justification for confessionalism's claims regarding the relationship between epistemology and method, the philosophy of science literature is useful to indicate the limits and possibilites for the further development of confessional theology along lines compatible with those claims.

On the one hand, progress within any field at any stage of its movement from undisciplined to fully disciplined inquiry can never be achieved through methodological debate alone. At best, the preoccupation of a field with methodological debate may be an accurate indicator of the fundamental unclarities that must be resolved before that area of study can achieve the status of disciplined inquiry. To that degree, preoccupation with methodology may be marginally preferable to the premature and therefore specious appearance of disciplined inquiry.

On the other hand, recent developments in epistemology point to where progress in theological inquiry needs to be sought: in the production of substantive results that win the approval of the community in whose name that inquiry is conducted. Progress in theology loyal to confessional method (or to any other method, for that matter) is achieved when the discussion of method is provisionally halted and the work of substantive theology is begun.

The success or failure of theological construction will be determined not by the debates of methodologists, but by the judgment of the Christian community. Those judgments are made when it assesses the product of such substantive work. It succeeds when that work properly expresses what that community collectively sees, albeit from a "limited point of view in history and faith." That community strives to interpret ever new possibilities that emerge in its ongoing historical encounter with "that great void . . . that enemy of all our causes . . . that enemy of all our Gods."[24] But in that encounter, the community confesses that it finds nothing other than ultimate reality, none besides the God of its confession.

NOTES

INTRODUCTION

1. Gordon Kaufman, *An Essay on Theological Method* (Missoula, Mont.: Scholars Press, 1975), ix. See also Douglas F. Ottati, *Meaning and Method in H. Richard Niebuhr's Theology* (Washington, D.C.: University Press of America, 1982), 171–98, for a lengthy discussion of this problem.

2. A very helpful analysis of the diversity of communities theologians address and the confusions it generates is found in David Tracy, *The Analogical Imagination: Christian Theology and the Culture of Pluralism* (New York: Crossroad, 1981), 3–46.

3. A concise survey of the diversity of contemporary theological methods and the kinds of decisions that theologians must make in developing and defending theological arguments is found in David H. Kelsey's excellent article, "Method, Theological," in *The Westminster Dictionary of Christian Theology,* ed. Alan Richardson and John Bowden (Philadelphia: Westminster Press, 1983), 363–68.

4. George A. Lindbeck, *The Nature of Doctrine: Theology in a Postliberal Age* (Philadelphia: Westminster Press, 1984).

5. As defined by H. Richard Niebuhr in *The Meaning of Revelation* (London: Macmillan, 1941). Another meaning—a theologian who uses confessional documents such as *The Book of Concord* as an authoritative source for theological construction—is not relevant here.

6. See Jeffrey Stout, *The Flight from Authority: Religion, Morality, and the Quest for Autonomy* (Notre Dame, Ind.: University of Notre Dame Press,

1981), 25–92, for a clear, helpful account of the cultural and historical origins of this way of framing the issues and of its impact on religious thought.

7. The idea of an "emerging consensus," of course, requires careful treatment. It is certainly not the case that philosophers as diverse as Stephen Toulmin, Thomas Kuhn, and Imre Lakatos, to name only a few, agree on questions of detail in the analysis of the development of scientific traditions. For example, Toulmin extensively criticizes Kuhn's notion of "scientific revolutions," as I will discuss below. Nevertheless, whatever their differences, these thinkers clearly bear a strong "family resemblance" to one another when compared with the earlier and now almost entirely abandoned attempt to formulate criteria for verification of meaning and to distinguish rigidly scientific and mathematical thought from all kinds of "meaningless" language which typified the philosophy of science of the Logical Positivist and Logical Empiricist periods.

8. Harold I. Brown, *Perception, Theory and Commitment: The New Philosophy of Science* (Chicago: University of Chicago Press, 1977), 10.

9. This revolution is commonly attributed to the first edition of Thomas Kuhn, *The Structure of Scientific Revolutions*, 2nd ed., enlarged (Chicago: University of Chicago Press, 1962), but reflects the consensus of a number of philosophers who reached quite similar conclusions independently.

10. The best exposition of such a rapprochement is in Richard Rorty, *Philosophy and the Mirror of Nature* (Princeton, N.J.: Princeton University Press, 1979).

11. See, for example, Richard Bernstein, *Beyond Objectivism and Relativism* (Philadelphia: University of Pennsylvania Press, 1985). See also Stout, *Flight from Authority*, 256ff.

12. The phrase, of course, is a reflection of Bernstein's *Beyond Objectivism*, but the fundamental idea is virtually a commonplace. See also Stout, *Flight from Authority*, for an extended discussion on this theme.

13. Lindbeck, *Nature of Doctrine*.

1: THE CONFESSIONAL NATURE OF SCIENCE

1. Many interesting details of this development cannot be treated here. But the discussion that follows owes a great deal to Harold I. Brown's excellent book, *Perception, Theory and Commitment: The New Philosophy of Science* (hereafter, *PTC*) (Chicago: University of Chicago Press, 1977), esp. 15–24.

2. Norwood Hanson, *Patterns of Discovery* (Cambridge: Cambridge University Press, 1958), 3.

3. Brown asserts, "The new approach to the philosophy of science has grown out of the failure of the older approach to solve its problems and out of anomalies revealed by modern studies of this history of science." *PTC*, 127.

4. For a lengthy discussion of the relationship between historical and philosophical analyses of science, see Thomas Kuhn, "The Relations between the History and the Philosophy of Science," in *The Essential Tension: Selected Studies in Scientific Traditions and Change* (Chicago: University of Chicago Press, 1977), 3–20.

5. C. I. Lewis, *An Analysis of Knowledge and Valuation* (La Salle, Ill.: Open Court, 1971), 171–72.

6. The term is originally Hanson's, but it has become part of the standard vocabulary of the discipline. See Hanson, *Patterns*, 54ff., for a full discussion of the concept, which he there calls the "theory loaded" character of words.

7. The use of gestalt shift experiment as an analog to the paradigms that guide a research discipline has been subjected to penetrating and correct criticism, noting especially that the gestalt figure cannot be extended or developed beyond the immediacy of the perceptual figures they generate, whereas the scientific paradigm is important precisely because it is subject to such extension. See Margaret Masterman, "The Nature of a Paradigm," in *Criticism and the Growth of Knowledge,* ed. Imre Lakatos and Alan Musgrave (New York: Cambridge University Press, 1970), 76. While imprecise, the use of the analogy of such figures dramatically communicates what Kuhn and others have in mind.

8. Stephen Toulmin, *Human Understanding* (hereafter, *HU*) (Princeton, N.J.: Princeton University Press, 1972), 150.

9. Thomas S. Kuhn, *The Structure of Scientific Revolutions*, 2nd ed., enlarged (hereafter, *SSR*) (Chicago: University of Chicago Press, 1962), 113.

10. This is only one of numerous examples. See *SSR*, 115–19. As he says in another context, "No part of the aim of normal science is to call forth new sorts of phenomena, indeed those that will not fit the box are often not seen at all" (*SSR*, 24.).

11. Notwithstanding that some of these accepted explanations are objectively "truer" than others, or superior to the others. See below, pp. 24–27.

12. For a discussion of this ambiguity, see Masterman, "Nature of a Paradigm," 59–89. There she identifies twenty-one distinct senses in which Kuhn uses the term.

13. See *SSR*, 25–27, for a full discussion of these three types.

14. Masterman, "Nature of a Paradigm," 73. See also 73–74 for her collection of Kuhn's scattered references to the problem of disciplinary formation and the rationale for her threefold division.

15. Ibid., 74.

16. Ibid. Although Kuhn does not have Masterman's clarity in distinguishing these differing situations, he does recognize the possibility of "local" progress within one school of a field that lacks an overall paradigm. The example he cites is that of Aristotelians among philosophers. See *SSR*, 162.

17. Masterman, "Nature of a Paradigm," 74.

18. Ibid.

19. Of course, although she does not note it, there are very significant differences as well, such as the fact that different astrologers given the same basic birth data do not formulate similar predictions. Still, the similarities are worth noting.

20. Ibid., 75.

21. The precise degree of incommensurability is a major matter of dispute within this tradition. Positions range from Kuhn's claim that such thinkers live "in different worlds" (*SSR*, 120ff.) to Toulmin's attempt to draw an analogy between judges reasoning from cases decided in jurisdictions of increasing distance from their own (*HU*, 85ff.). See the discussion of the problem of relativism below (pp. 24–27).

22. *SSR*, 76. See the discussion here for other examples of the nonrational factors that have affected or determined the acceptance or rejection of new paradigms in other sciences.

23. *SSR*, 152–59. Kuhn proceeds in this section to discuss individual idiosyncrasy, nationality, aesthetic appeal, and collegial relations as other nonrational factors affecting paradigm acceptance.

24. See *PTC*, 121–26, for the details of this analysis. See also *HU*, 148ff., for a discussion of the continuity of a discipline through the "genealogy of problems" that are addressed by a disciplinary group over time.

25. *PTC*, 121. This is only one of numerous examples that Brown develops of the closeness of the link between the meaning of a concept and its location within and relation to other concepts within a theory.

26. Richard Rorty, *Philosophy and the Mirror of Nature* (Princeton, N.J.: Princeton University Press, 1979), 267. See the entire book for a full and provocative discussion of the epistemology presupposed by this "mirroring" metaphor, and of an alternative epistemology which he feels must replace it. See also Jeffrey Stout, *The Flight from Authority: Religion, Morality, and the Quest for Autonomy* (Notre Dame, Ind.: University of Notre Dame Press, 1981), 256–72, for a clear and subtle statement of the relation between historicism and truth claims.

27. See *SSR*, 170ff. For a further discussion of the evolutionary analogy, see Stephen Toulmin, "Does the Distinction between Normal and Revolutionary Science Hold Water?" in *Criticism and the Growth of Knowledge*, ed.

Lakatos and Musgrave, 39–50. There Toulmin is concerned to extend the evolutionary analogy and to use it to criticize what he considers to be Kuhn's excessive "catastrophism" in his account of scientific "revolutions."

28. While not elaborated to any great extent, the core of this idea is found in *PTC*, 94.

2: THE SPECTRUM OF CONFESSIONAL THEOLOGIES

1. Karl Barth, *Church Dogmatics* (hereafter, *CD*) (Edinburgh: T. and T. Clark, 1975), 1.1:12. See also *CD*, 2.2:509–43, where a similar point, the autonomy of theological ethics with reference to philosophical ethics, is argued at great length. Typical of this discussion is the following: "We must first refuse to follow all those attempts at theological ethics which start from the assumption that it is to be built on, or to proceed from, a general human ethics, a 'philosophical' ethics" (543).

2. *CD*, 1.1:30. Note again the stress on the contingency and fallibility of any given theological construction.

3. *CD*, 1.1:42. See also *CD*, 2.2:71–132. Here Barth attacks the notion of beginning with an anthropology based in the sciences to determine the essential nature of humans. Instead, "the man Jesus" must be taken as the proper ground for anthropology, since he alone is "real man."

4. Of course, how other sciences could accept this criterion in practice is not clear (which Barth would, of course, point to as further evidence of our separation from God). For our purposes here, however, that issue is less important than the comparative point made with reference to Holmer and Phillips, for whom properly religious language is necessarily separated from other spheres of discourse. These passages at least indicate how vigorously Barth wishes to deny that claim.

5. The phrase is Holmer's. He claims precisely the opposite (see below, p. 37).

6. Holmer himself seems to recognize the closeness of his position to an apologia for anti-intellectualism in theology. He feels obligated (having said so much to suggest that interpretation) to assert that he is not offering a defense for "plebian piety." See "Theology, Atheism, and Theism," (hereafter, TAT), in his collection of essays, *The Grammar of Faith* (hereafter, *GF*) (New York: Harper & Row, 1978), 167.

7. Paul Holmer, "What Theology Is and Does—Again" in *GF*, 30–31.

8. Paul Holmer, "Scientific Language and the Language of Religion" (hereafter, SLLR), in *GF*, 54–80.

NOTES

9. This contrast is superficially similar, of course, to that drawn by Niebuhr between "inner" and "outer" history. The distinction between them is that Niebuhr is quite clear that connections between the two kinds of history are common whereas Holmer, in his moments of greatest clarity, explicitly denies the very possibility of points of contact between them.

10. WTID, 11. Of course, this passage lumps together many different things and dismisses them with one stroke. It is one thing to understand a Christian theology as an intellectual proposal which (leaving out the pejorative "assimilation of novelties") adapts to new circumstances. It is something else entirely, perhaps related but not obviously or necessarily so, to think that such proposals will "reclaim the masses." Intelligibility and persuasion are not the same thing.

11. Paul Holmer, "Metaphysics and Theology" (hereafter, MT) in *GF,* 92–93.

12. Paul Holmer, "Theology and Concepts" (hereafter, TC) in *GF,* 149.

13. Paul Holmer, "Language and Theology" in *GF,* 133–34. See also SLLR, 73, where Holmer asserts that religious and moral judgments are "about everything else in the world"—about "the large cosmos in addition to ourselves."

14. This is potentially obscured by his need to distinguish in the course of his analysis between "genuine" religious utterances and other utterances of believers which are religious "blunders."

15. D. Z. Phillips, "Faith, Scepticism, and Religious Understanding," in *Faith and Philosophical Enquiry* (hereafter, *FPE*) (New York: Schocken Books, 1970), 17.

16. D. Z. Phillips, "Religious Beliefs and Language-Games" (hereafter, RBLG), in *FPE,* 82.

17. An example of the function of such a world-picture is the idea of a Resurrection Day and Judgment in the moral life of a believer. Rather than being a belief in a future event analogous to, for example, the belief in 1987 that George Bush would win the 1988 Presidential election, this belief "might play the role of constantly admonishing me, or I always think of it. Here, there is an enormous difference between those people for whom the picture is constantly in the foreground, and the others who just didn't use it at all. Those who said: 'Well, possibly it may happen and possibly not' would be on an entirely different plane.' " Ludwig Wittgenstein, *Lectures and Conversations on Aesthetics, Psychology, and Religious Belief* (Berkeley: University of California Press, 1972), 56.

18. See D. Z. Phillips, "Does God Exist?" (hereafter, DGE), in *Religion Without Explanation* (hereafter, *RWE*) (Oxford: Basil Blackwell, 1976), 162–63.

NOTES

19. D. Z. Phillips, "Philosophy, Theology and the Reality of God" (hereafter, PTRG), in *FPE*, 4.

20. For example, there is no method of determining that conjugating a particular verb according to a standard paradigm will be a blunder because the verb is irregular. Only the usage of native speakers can show which verbs are regular and which irregular.

21. D. Z. Phillips, "Unconscious Reductionism," in *RWE*, 148–49.

22. John Passmore, "Christianity and Positivism," *Australasian Journal of Philosophy* 35(1957): 128, as quoted in RBLG, 93.

23. D. Z. Phillips, *The Concept of Prayer* (hereafter, *CP*) (London: Routledge and Kegan Paul, 1965), 1.

24. D. Z. Phillips, "Faith, Scepticism, and Religious Understanding," in *FPE*, 23.

25. Ibid., 23–24.

26. The very fact that we distinguish linguistically between "religion" and "superstition," Phillips argues, indicates our awareness of the need to mark the difference. See D. Z. Phillips, "Religion, Magic, and Metaphysics," in *RWE*, 111ff.

27. For a fuller discussion of this problem, however, see Alan Keightley, *Wittgenstein, Grammar and God* (London: Epworth Press, 1976), 67–70. Generally, Keightley's treatment of Wittgenstein and his successors is an excellent introduction to the problems and promise of the Wittgensteinian tradition in philosophy of religion.

28. George Lindbeck, *The Nature of Doctrine: Theology in a Postliberal Age* (hereafter, *ND*) (Philadelphia: Westminster Press, 1984).

29. As David Tracy puts it, "In a sense [Lindbeck] has written two books. One study . . . is on the nature of doctrine. . . . The second study . . . is yet more ambitious. Here Lindbeck articulates a new paradigm . . . geared to understanding religions as analogous to languages and cultures as well as to understanding theologies as largely grammatical enterprises." "Lindbeck's New Program for Theology: A Reflection," *The Thomist* 49(1985): 460.

30. *ND*, 16. In addition, there are "hybrid approaches," which attempt to combine both emphases, but these can be subsumed under the other two for the purpose of analysis of their merit, he asserts.

31. Indeed, Lindbeck carefully qualifies the kind of argument that is *possible* between fundamentally different perspectives on religion and religious thought. *ND*, 10–11.

32. But again, see Tracy for significant qualification, at least, of Lindbeck's claims regarding experiential-expressivist methods on this point, "Lindbeck's New Program," 462–64. Lindbeck himself is a bit more subtle on this point later in the book: "Turning now . . . to the relation of religion and experience,

NOTES

it may be noted that this is not unilateral but dialectical. It is simplistic to say (as I earlier did) merely that religions produce experiences, for the causality is reciprocal" (*ND*, 33).

33. One way to cope with this problem is, of course, advocating a fully sectarian stance of cultural withdrawal from the din of competing perspectives that constitute modern life. Lindbeck himself is aware of the sociological implications of his line of argument (*ND*, 133). This set of problems will be taken up again in the conclusion (p. 108).

34. Richard John Neuhaus, "Is There Theological Life after Liberalism: The Lindbeck Proposal," *Dialog: A Journal of Theology* 24 (Winter 1985): 70.

35. Tracy, "Lindbeck's New Program," 465.

36. Ibid., 469–70.

37. Ibid., 470.

38. Francis Schüssler Fiorenza, *Foundational Theology: Jesus and the Church* (New York: Crossroad, 1984). A concise and somewhat less technical presentation of his position is available in his 1986 Presidential Address to the Catholic Theological Society of America, "Foundations of Theology: A Community's Tradition of Discourse and Practice," *CTSA Proceedings* 41(1986): 107–34.

39. *FT*, 273–74. Despite his criticisms of an interpretation of the nature of theological truth claims exclusively in terms of correspondence to reality, Fiorenza is also careful to agree with Pannenberg's realism that theological claims must be "about something." On the other hand, Fiorenza argues that Pannenberg is also one-sided in his failure to emphasize the point that religious truth is "more than facticity" and that there is a distinctively religious dimension to such truth which is not exhausted by its factual content (*FT*, 274).

40. See *FT*, 276ff., for a discussion of various correlation methods.

3: H. RICHARD NIEBUHR'S NONINTERNALIST PROPOSAL

1. H. Richard Niebuhr, *The Responsible Self* (hereafter, *RS*) (New York: Harper & Row, 1963), 44–45.

2. This point will be elaborated in some detail in a later section of this chapter, where it is extremely important in understanding Niebuhr's defense against the charge that confessional theology is necessarily committed to "internalism."

3. H. Richard Niebuhr, *The Meaning of Revelation* (hereafter, *MR*) (New York: Macmillan, 1941).

4. A typology similar in many ways to Lindbeck's.

116

NOTES

5. H. Richard Niebuhr, "Value Theory and Theology" (hereafter, VTT), in *The Nature of Religious Experience* (New York: Harper & Brothers, 1937), 97.

6. Cf. H. Richard Niebuhr, "The Center of Value," in R. N. Anshen, ed., *Moral Principles of Action* (New York: Harper & Brothers, 1952), 162–75. Also printed in H. Richard Niebuhr, *Radical Monotheism and Western Culture with Supplementary Essays* (hereafter, *RM*) (New York: Harper & Row, 1970),100–13.

7. VTT, 111. Niebuhr's willingness to grant the necessity of a general theory of religion, revelation, and religious ethics in order to justify assertion and denial significantly distinguishes him from the Wittgensteinian theologians discussed in chap. 1. Since this point bears most directly on the question of internalism in Niebuhr, it will be explored more fully later in this chapter.

8. See e.g., "Radical Monotheism and Western Religion," in *RM*, 60.

9. H. Richard Niebuhr, "The Triad of Faith," in *Andover-Newton Bulletin* 47 (October 1954): 6–7.

10. For the classic example of Niebuhr's attempt to apply a pattern derived from the particularity of the "inner" history of the Christian community to new and uninterpreted experience, see his war articles, "War as the Judgment of God," *The Christian Century* 59 (May 1942): 630–33; "Is God in the War?" *The Christian Century* 59 (August 1942): 953–55; and "War as Crucifixion," *The Christian Century* 60 (April 1943): 513–15.

11. There are, of course, enormous difficulties in being precise regarding a method by which this might be done, and ethical dangers involved in the misapplication of symbols from the tradition to historical experience. Niebuhr himself is aware of this when he considers and rejects the image of retribution as an adequate interpretation of World War II in the article "War as Crucifixion." I will not explore these difficulties, however, since they lie beyond the issue at hand: the demonstration that Niebuhr cannot fairly be charged with internalism.

4: AN OPEN CONFESSIONAL METHOD

1. One must not overdo this analogy between theology and an undisciplined field, since it also is the case that it is a "field-encompassing field," even at its best. I am indebted to Langdon Gilkey's remarks on an earlier draft of this chapter for this clarification.

2. See Brevard Childs, *Biblical Theology in Crisis* (Philadelphia: Westminster Press, 1970), for a lengthy discussion of the rise and fall of biblical theology. Note in particular his close analysis of the ways in which the collapse of the

movement was precipitated by internal incoherence in the assumptions that governed the movement.

3. See Margaret Masterman, "The Nature of a Paradigm," in *Criticism and the Growth of Knowledge,* ed. Imre Lakatos and Alan Musgrave (New York: Cambridge University Press, 1970), 75, for a discussion of premature setting-in of paradigm-guided inquiry. Of course, even in such comparatively rare periods in modern theology, the degree of unanimity among thinkers sharing that orientation is substantially looser than in a fully disciplined scientific field. Also, it is valuable to reflect on whether such periods of general methodological agreement are not necessarily episodic given the nature of theology in contrast to the long-term cumulative efforts characteristic of scientific disciplines. Nevertheless, as we saw in the discussion of continuity and discontinuity within scientific disciplines, it is important not to overemphasize continuity based solely on the persistence of terms within a discipline.

4. H. Richard Niebuhr, *The Meaning of Revelation* (hereafter, *MR*) (New York: Macmillan, 1960), 29.

5. See Thomas Byrnes, "H. Richard Niebuhr's Christian Moral Philosophy" (Ph.D. diss., University of Chicago, 1982), chap. 1, passim, for a full discussion of the relation between Niebuhr's position and Troeltsch's.

6. It is important to remember (although critics of confessional method often fail to) that the claim that theology must have such a starting point is logically distinct from the claim that theology must remain throughout its work "inside" a given community or perspective (the claim regarding internalism). Confessional theologians are united in making some version of the claim regarding the starting point but, as demonstrated in chap. 2, not all confessional positions are internalist.

7. See David Tracy, for example, for a recognition of the complexity of the relationship between experience and language even from the side of a proponent of and approach to theology which attempts to begin with "common human experience." "Lindbeck's New Program for Theology," *The Thomist* 49 (1985): 463–65.

8. See again Masterman, "Nature of Paradigm," 73–75 for a clear discussion of this predisciplinary phase.

9. H. Richard Niebuhr, *The Responsible Self* (hereafter, *RS*) (New York: Harper & Row, 1963), 79–80.

10. Examples of such methodological preoccupation are legion. To name only a couple of examples, see Gordon Kaufman, *An Essay in Theological Method* (Missoula, Mont.: Scholars Press, 1975); and David Tracy, *Blessed Rage for Order: The New Pluralism in Theology* (New York: Seabury Press, 1975). Tracy is distinctive among those intensely engaged in methodological inquiry, however, in that he essays to move beyond methodological argumentation to

the contribution of a substantive theological construction. See also his *The Analogical Imagination: Christian Theology and the Culture of Pluralism* (New York: Crossroad, 1981) for the constructive application of his methodological argumentation.

11. I hasten to add, of course, that the notion of "assured results" in theology must necessarily be considerably looser than in the sciences, where mathematical precision is often possible. Nevertheless, one may reasonably extend the notion to include at least agreement on fundamental perspectives, materials, and so forth—a goal that theology has attained during periods in the past, and can reasonably aspire to even now.

12. Marx and Feuerbach are only two of several examples he gives here, but unfortunately all are merely examples. There is no general characterization of the kinds of external perspectives these are and why they are appropriately incorporated. Nor are there any examples of external perspectives which are not to be incorporated from which one might generate some provisional characterization of the distinction between the two groups.

13. Thomas S. Kuhn, *The Structure of Scientific Revolutions* (hereafter, *SSR*) (Chicago: University of Chicago Press, 1962), 152.

14. Stephen Toulmin, *Human Understanding* (hereafter, *HU*) (Princeton, N. J.: Princeton University Press, 1972), 151.

15. As I mentioned in the detailed analysis of Niebuhr's view earlier, the best example of Niebuhr's attempt to do precisely this task is found in his famous "War articles" (see chap. 3, n. 10) in which he attempts to explain the events of World War II as manifestations of the purposes of the God known to Christian faith. There, he weighs the relative merits of several explanatory models provided by the Christian tradition (such as punishment, retribution, judgment) and then argues for the relative adequacy of crucifixion as the best among them for accounting for the salient features of the experience of that war.

16. On the nature of such limit questions, see Tracy, *Blessed Rage for Order*, 92–109.

17. For an instructive example of such diversity, see the centrality of the literary model of "the Classic" for the development of systematic theology in Tracy, *Analogical Imagination*, and compare it to the centrality of modern scientific cosmology in theological reformulation in James Gustafson, *Ethics from a Theocentric Perspective* (Chicago: University of Chicago Press, 1981), vol. 1. Such an example clearly illustrates, as the work of practically any pair of modern theologians would, the fact that there is no agreement on which fields and disciplines beyond the traditional theological ones are essential to contemporary constructive theology.

18. Again, however, this is not to suggest that one overlook the quite significant difference that scientific disciplines are inherently superior to theology in any form in their ability to more precisely delineate circles of greater and lesser relevance and to exclude from their range of concern aspects of experience which the understanding of the field at any given point in its history determines to fall outside the bounds of appropriate disciplinary concern.

19. This is a point, however, on which it is important not to confuse Niebuhr's position with that of some contemporary theologians of "story" who find some of their inspiration in his work. While the story for Niebuhr is constantly a point of return and locus of critical comparison of more speculative theological formulations, he nevertheless recognizes the legitimacy and importance of those formulations in a way often obscured if not denied by many of the theology of story advocates. See, for example, Stanley Hauerwas, "Story and Theology," in *Truthfulness and Tragedy: Further Investigations into Christian Ethics* (Notre Dame, Ind.: University of Notre Dame Press, 1977), 71–81. Hauerwas writes, "For like the self, God is a particular agent that can be known only as we know his story" (79), which from Niebuhr's perspective confuses the epistemological limitations on our ability to think and speak of God within a common framework.

20. D. Z. Phillips writes, "the attention of the individual has been won over either by a rival secular picture, or, of course, by worldliness, etc." ("Religious Beliefs and Language Games," in *Faith and Philosophical Enquiry* [New York: Schocken Books, 1970], 115). Elsewhere, discussing the question of those who do not share a particular religious framework of discourse, preferring another framework, he writes, "These are genuine alternatives since they indicate that the person has no use for the religious belief, that it means nothing to him, that he does not live by such a belief, or that he holds other beliefs which exclude religious faith. In this latter case, however, the alternatives are not alternatives with the same mode of discourse, but rather, different perspectives on life" ("Does God Exist?" in his *Religion without Explanation* [Oxford: Basil Blackwell, 1976], 6).

21. The phrase is William A. Christian's. For an analysis of the logic of religious truth claims which is highly congenial to mine, see his *Meaning and Truth in Religion* (Princeton, N.J.: Princeton University Press, 1964).

22. H. Richard Niebuhr, "Value Theory and Theology," in *The Nature of Religious Experience* (New York: Harper & Brothers, 1937), 105.

23. For example, see Gustafson, *Ethics from a Theocentric Perspective*, 1:225–35. His much earlier *Treasure in Earthen Vessels: The Church as a Human Community* (New York: Harper & Row, 1961) is an extensive treatment of the nature of the church from a sociological perspective highly congenial to

Niebuhr's approach. Gordon Kaufman's early *Systematic Theology: A Historicist Perspective* (New York: Charles Scribner's Sons, 1968) is an attempt to do confessional theology from a Niebuhrian perspective. His much later work, *An Essay on Theological Method* (Missoula, Mont.: Scholars Press, 1975), while superficially far removed from confessional method, can arguably be seen as an extension of confessional method in the absence of a cohesive religious community to the "community" of Western culture and the meaning it has assigned to the word "god."

24. H. Richard Niebuhr, *Radical Monotheism and Western Culture with Supplementary Essays* (New York: Harper & Row, 1970), 122–23.

BIBLIOGRAPHY

I. BOOKS

Armstrong, D. M. *Belief, Truth, and Knowledge*. London and New York: Cambridge University Press, 1973.

Barth, Karl. *Church Dogmatics*. Edited by G. W. Bromiley and T. F. Torrance. Translated by G. W. Bromiley et. al. Edinburgh: T. & T. Clark, 1957.

Bernstein, Richard. *Beyond Objectivism and Relativism: Science, Hermeneutics, and Praxis*. Philadelphia: University of Pennsylvania Press, 1985.

Brown, Harold I. *Perception, Theory and Commitment*. Chicago: University of Chicago Press, 1977.

Brown, Stuart C., ed. *Reason and Religion*. Ithaca and London: Cornell University Press, 1977.

Christian, William A. *Oppositions of Religious Doctrines*. New York: Herder & Herder, 1972.

Christian, William A. *Meaning and Truth in Religion*. Princeton, N.J.: Princeton University Press, 1964.

Crosson, Frederick, ed. *The Autonomy of Religious Belief*. University of Notre Dame Studies in the Philosophy of Religion, vol. 2. Notre Dame, Ind.: University of Notre Dame Press, 1981.

Delaney, C. F., ed. *Rationality and Religious Belief*. University of Notre Dame Studies in the Philosophy of Religion, vol. 1. Notre Dame, Ind.: University of Notre Dame Press, 1979.

Fiorenza, Francis Schüssler. *Foundational Theology: Jesus and the Church*. New York: Crossroad, 1984.

BIBLIOGRAPHY

Fowler, James W. *To See the Kingdom: The Theological Vision of H. Richard Niebuhr.* Nashville: Abingdon Press, 1974.

Gustafson, James M. *Ethics from a Theocentric Perspective: Theology and Ethics.* Vol. 1. Chicago: University of Chicago Press, 1981.

Gustafson, James M. *Treasure in Earthen Vessels: The Church as a Human Community.* New York: Harper & Row, 1961.

Hanson, Norwood Russell. *Patterns of Discovery.* Cambridge: Cambridge University Press, 1958.

Hartt, Julian N. *Theological Method and Imagination.* New York: Seabury Press, 1977.

Harvey, Van A. *The Historian and the Believer.* New York: Macmillan Company, 1966.

Hoedemaker, Libertus A. *The Theology of H. Richard Niebuhr.* Philadelphia: Pilgrim Press, 1970.

Holmer, Paul L. *The Grammar of Faith.* New York: Harper & Row, 1978.

Hook, Sidney. *Religious Experience and Truth.* New York: New York University Press, 1961.

Hudson, Donald. *Ludwig Wittgenstein.* Makers of Contemporary Theology. Richmond, Va.: John Knox Press, 1968.

Irish, Jerry Arthur. *Revelation in the Theology of H. Richard Niebuhr.* Ann Arbor, Mich.: University Microfilms, 1982.

————. *The Religious Thought of H. Richard Niebuhr.* Atlanta: John Knox Press, 1983.

Kaufman, Gordon D. *An Essay on Theological Method.* AAR Studies in Religion, vol. 11. Missoula, Mont.: Scholars Press, 1975.

————. *God the Problem.* Cambridge, Mass.: Harvard University Press, 1972.

————. *Systematic Theology: A Historicist Perspective.* New York: Charles Scribner's Sons, 1968.

————. *The Theological Imagination.* Philadelphia: Westminster Press, 1981.

————. *Relativism, Knowledge, and Faith.* Chicago: University of Chicago Press, 1969.

Keightley, Alan. *Wittgenstein, Grammar and God.* London: Epworth Press, 1976.

Kenney, Alan. *Wittgenstein.* Cambridge, Mass.: Harvard University Press, 1973.

Klemke, E. D., ed. *Essays on Wittgenstein.* Urbana, Ill.: University of Illinois Press, 1971.

Kliever, Lonnie D. *H. Richard Niebuhr.* Makers of the Modern Theological Mind. Waco, Texas: Word Incorporated, 1977.

Kuhn, Thomas S. *The Copernican Revolution.* Cambridge, Mass.: Harvard University Press, 1957.

————. *The Essential Tension: Selected Studies in Scientific Traditions and Change*. Chicago: University of Chicago Press, 1977.

————. *The Structure of Scientific Revolutions*. International Encyclopedia of Unified Science, vol. 2 no. 2. 2d ed., enlarged. Chicago: University of Chicago Press, 1962.

Lakatos, Imre, and Alan Musgrave, eds. *Criticism and the Growth of Knowledge*. Proceedings of the International Colloquium in Philosophy of Science, vol. 4. New York: Cambridge University Press, 1970.

✓ Lindbeck, George A. *The Nature of Doctrine: Religion and Theology in a Postliberal Age*. Philadelphia: Westminster Press, 1984.

Mead, George H. *Mind, Self, and Society*. Chicago: University of Chicago Press, 1934.

Morawetz, Thomas. *Wittgenstein & Knowledge*. Amherst: University of Massachusetts Press, 1978.

Neville, Robert C. *Reconstruction of Thinking*. Albany: State University of New York, 1981.

Niebuhr, H. Richard. *The Meaning of Revelation*. New York: Macmillan Co., 1941.

————. *Christ and Culture*. New York: Harper & Row, 1951.

————. *The Kingdom of God in America*. New York: Harper & Row, 1937.

————. *The Purpose of the Church and Its Ministry*. Harper Ministers Paperback Library, no. RD 211. New York: Harper & Row, 1956.

————. *Radical Monotheism and Western Culture*. New York: Harper & Brothers, 1960.

————. *The Responsible Self*. New York: Harper & Row, 1963.

————. *The Social Sources of Denominationalism*. New York: The World Publishing Company, 1929.

Ottati, Douglas F. *Meaning and Method in H. Richard Niebuhr's Theology*. Washington, D.C.: University Press of America, 1982.

Penelhum, Terence. *Problems of Religious Knowledge*. New York: Herder & Herder, 1972.

Pepper, Stephen C. *World Hypotheses*. Berkeley and Los Angeles: University of California Press, 1942.

Phillips, D. Z. *The Concept of Prayer*. London: Routledge & Kegan Paul, 1965.

————. *Faith and Philosophical Enquiry*. New York: Schocken Books, 1970.

————. *Through a Darkening Glass*. Notre Dame, Ind.: University of Notre Dame Press, 1982.

————. *Death and Immortality*. London: MacMillan & Co. 1970.

————. *Religion Without Explanation*. Oxford: Basil Blackwell, 1976.

Pitcher, George, ed. *Wittgenstein: The Philosophical Investigations*. Notre Dame, Ind.: University of Notre Dame Press, 1968.

BIBLIOGRAPHY

Polanyi, Michael. *Personal Knowledge: Toward a Post-Critical Philosophy*. Corrected ed. Chicago: University of Chicago Press, 1962.

Popper, Karl R. *The Poverty of Historicism*. New York: Harper & Row, 1957.

———. *The Logic of Scientific Discovery*. New York: Harper & Row, 1959.

———. *Objective Knowledge: An Evolutionary Approach*. London: Oxford University Press, 1972.

Quine, Willard V. O. *Ontological Relativity and Other Essays*. The John Dewey Essays in Philosophy, vol. 1. New York: Columbia University Press, 1969.

———. *Word & Object*. Cambridge: MIT Press, 1960.

———. *From a Logical Point of View*. 2d ed., rev. New York: Harper & Row, 1961.

Quine, Willard V. O., and J. S. Ullian, *The Web of Belief*. 2d ed. New York: Random House, 1970.

Ramsey, Paul, ed. *Faith and Ethics: The Theology of H. Richard Niebuhr*. New York: Harper & Row, 1957.

Rescher, Nicholas. *The Primacy of Practice*. Oxford: Basil Blackwell, 1973.

Rorty, Richard. *Philosophy and the Mirror of Nature*. Princeton, N.J.: Princeton University Press, 1979.

Salmon, Wesley C. *The Foundations of Scientific Inference*. Pittsburgh: University of Pittsburgh Press, 1966.

Santoni, Ronald E., ed. *Religious Language and the Problem of Religious Knowledge*. Bloomington, Ind., and London: Indiana University Press, 1968.

Stout, Jeffrey. *The Flight from Authority: Religion, Morality, and the Quest for Autonomy*. Notre Dame, Ind.: University of Notre Dame Press, 1981.

———. *Ethics After Babel: The Languages of Morals and Their Discontents*. Boston: Beacon Press, 1988.

Toulmin, Stephen. *The Philosophy of Science: An Introduction*. Harper Torchbooks/The Science Library. New York: Harper & Row, 1960.

———. *Human Understanding*. Vol. 1. Princeton, N.J.: Princeton University Press, 1972.

———. *The Return to Cosmology*. Berkeley, Los Angeles, London: University of California Press, 1982.

Trigg, Roger. *Reason and Commitment*. London and New York: Cambridge University Press, 1973.

Williams, Michael. *Groundless Belief: An Essay in the Possibility of Epistemology*. Library of Philosophy and Logic. New Haven, Conn.: Yale University Press, 1977.

Wisdom, John. *Philosophy and Psychoanalysis*. Berkeley and Los Angeles: University of California Press, 1969.

Wittaker, John H. *Matters of Faith and Matters of Principle*. San Antonio: Trinity University Press, 1981.

BIBLIOGRAPHY

Wittgenstein, Ludwig. *Philosophical Investigations*. Translated by G. E. M. Anscombe. New York: Macmillan Company, 1953.

————. *The Blue and Brown Books*. New York: Harper & Row, 1958.

————. *Philosophical Investigations*. 3d ed. Translated by G. E. M. Anscombe. New York: Macmillan Company, 1958.

————. *Notebooks, 1914–1916*. Translated by G. E. M. Anscombe. New York: Harper & Row, 1961.

————. *Lectures and Conversations on Aesthetics, Psychology, and Religious Belief*. Edited by Cyril Barrett. Berkeley: University of California Press, 1972.

————. *Tractatus Logico-Philosophicus*. Translated by D. F. Pears and B. F. McGuinness. International Library of Philosophy and Scientific Method. London: Routledge & Kegan Paul, 1961.

————. *On Certainty*. Edited by G. E. M. Anscombe and G. H. von Wright. Translated by Denis Paul and G. E. M. Anscombe. New York: Harper & Row, 1969.

II. ARTICLES BY H. RICHARD NIEBUHR

"Back to Benedict?" *The Christian Century* 42 (1925): 860–61.

"What Holds Churches Together?" *The Christian Century* 43 (1926): 346–48.

"Churches That Might Unite." *The Christian Century* 56 (1929): 259–61.

"The Grace of Doing Nothing." *The Christian Century* 49 (1932): 378–80.

"What Then Must We Do?" *The Christian Century Pulpit* 5 (1934): 145–47.

"Man the Sinner." *Journal of Religion* 15 (1935): 272ff.

"Value Theory and Theology." In *The Nature of Religious Experience: Essays in Honor of D. C. Macintosh*, pp. 93–116. New York: Harper & Brothers, 1937.

"War as the Judgment of God." *The Christian Century* 59 (May 1942): 630–33.

"Is God in the War?" *The Christian Century* 59 (August 1942): 953–55.

"War as Crucifixion." *The Christian Century* 60 (April 1943): 513–15.

"Towards a New Other-Worldliness." *Theology Today* 1 (1944): 78–87.

"The Ego-Alter Dialectic and the Conscience." *Journal of Philosophy* 42 (1945): 352–59.

"The Norm of the Church." *Journal of Religious Thought* 4 (1946): 5–15.

"The Doctrine of the Trinity and the Unity of the Church." *Theology Today* 3 (1946): 371–84.

"Utilitarian Christianity." *Christianity and Crisis* 6, no. 12 (1946): 3–5.

"The Gift of Catholic Vision." *Theology Today* 4 (1948): 507–21.

"The Triad of Faith." *Andover-Newton Bulletin* 47 (1954): 3–12.

BIBLIOGRAPHY

"Theology—Not Queen but Servant." *Journal of Religion* 35 (1955): 1–5.
"Reformation, Continuing Imperative." *The Christian Century* 77 (Mar. 2, 1960): 248–51.
"On the Nature of Faith." In Sidney Hook, ed., *Religious Experience and Truth*, pp. 93–102. New York: New York University Press, 1961.
"How My Mind Has Changed." In H. E. Fey, ed., *How My Mind Has Changed*, pp. 69–90. Meridian Books. Cleveland: World Publishing Co., 1961.

III. OTHER ARTICLES

Buckley, James. "Doctrine in the Diaspora." *The Thomist* 49 (1985): 443–59.
Fiorenza, Francis Schüssler. "Foundations of Theology: A Community's Tradition of Discourse and Practice." Presidential Address. *CTSA Proceedings* 41 (1986): 107–34.
Jackson, Timothy P. Review of George A. Lindbeck, *The Nature of Doctrine*. *Religious Studies Review* 11 (July, 1985): 240–45.
Neuhaus, Richard John. "Is There Theological Life after Liberalism: The Lindbeck Proposal." *Dialog: A Journal of Theology* 24 (Winter 1985): 66–72.
O'Neill, Colman. "The Rule Theory of Doctrine and Propositional Faith." *The Thomist* 49 (1985): 417–42.
Placher, William. "Revisionist and Postliberal Theologies and the Public Character of Theology." *The Thomist* 49 (1985): 392–416.
Tracy, David. "Lindbeck's New Program for Theology: A Reflection." *The Thomist* 49 (1985): 460–72.
Wood, Charles M. Review of George A. Lindbeck, *The Nature of Doctrine*. *Religious Studies Review* 11 (July 1985): 235–40.

AUTHOR INDEX